The Lachine Canal

YVON DESLOGES
ALAIN GELLY

The Lachine Canal

Riding the Waves of Industrial and Urban Development

1860-1950

Translation by Donald Kellough
Revision by Jane Macauley

SEPTENTRION

Les éditions du Septentrion wishes to thank the Canada Council for the Arts and the Société de développement des entreprises culturelles du Québec (SODEC) for support of its publishing program, as well as the Government of Quebec for its tax credit program for book publishing. We are also grateful for financial support received from the Government of Canada through the Book Publishing Industry Development Program (BPIDP).

Copy editing: Catherine Broué and Joan Irving

Graphic Design: Folio infographie

Cover page photos: *General View, Montreal Harbour, Looking Eastward,* Centre d'archives d'Hydro-Québec, Shawinigan Water and Power Co. collection, F1 / 700862, 1931; *Old Lock Number Two, Looking East,* National Archives of Canada, Department of public Works Collection, PA 110242, 1908.

Back page photos: *Youth Diving into the Lachine Canal,* St. Henri Historical Society, 23.1-5.4, ca 1940; *Gabrielle Roy and Boys near St. Ambroise Street Crossing,* Archives nationales du Québec, Conrad Poirier collection, P48, P11917, 1945.

Published in collaboration with Parks Canada.

Les éditions du Septentrion, 2002
1300, ave Maguire
Sillery (Québec) G1T 1Z3
www.septentrion.qc.ca

Diffusion in Canada:
Diffusion Dimedia
539, boul. Lebeau
Saint-Laurent (Québec) H4N 1S2

Legal deposit – 3rd quarter 2002
National Library of Canada
ISBN 2-89448-331-7

Table of Contents

Foreword 9

Introduction 11

The Era of Canals 17
The British "Canal Mania" 19
The American "Canal Era" 20
The St. Lawrence Canals 20
The Lachine Canal and the Surrounding Industrial Complex 22
A Bird's-eye View 25
The Basic Principles and Infrastructures of Canals 27
Ocean Basins 32
Basins Reserved for Inland Fleets 38
Around the Locks 42
The Ville LaSalle – Ville Saint-Pierre Sector 45
Lachine 49
Workers of the Canal 55
Floods 61

Shipping and Fleets 63
Bulk Shipment 65
Shipping Lines 66
The Harbour Landscape 67
Transported from the 19th into the 20th Century 69
Canal Fleets 71
Shipping Regulations 79
Shipowners and Shipping Lines 90
Shipbuilding and Repairing 93
Bulk Transport 96
Passenger Transportation 101
Workforce and Infrastructures: Unloading Cargo 105

The road from Hydraulic Power to Electricity— by Way of Smoky Valley 109
The Rise of a Hydraulic Complex 111
The Beginnings of Smoky Valley 112
The Development of the Lachine Canal Industrial Corridor Continues Apace 113
The Era of Large Companies 114
Energy 117
Hydraulic Power 119
Coal and Steam 125
Electricity 129
Industrial Star Performers 133
The Consumer Goods Industry 135
Heavy Industry 146
Chemicals and Carbochemical Products 155
The Electrical Industry 158

Working in the Factory, Living in the City 161
Work in the Factory: from Tolerating the Intolerable to the Quest fcr quality of Life 163
City Living 166
The World of Work 169
Working Conditions 171
Working Hours 178
Working Women 181
Child Labour 186
Urbanization of the Corridor 189
Working-class Communities 191
Urban Planning 194
Housing Conditions 199
Urban Mass Transit 204
Daily Life around the Canal 207

Epilogue 209

Select Bibliography 213

Foreword

The Lachine rapids, which the French originally called Sault Saint Louis, explain Montreal. Cities often spring up where there is an obstacle—a mountain chain, a waterway, or some hindrance to navigation. As the traveller moves upriver from the ocean, the St. Lawrence presents no falls or dangerous rapids for more than 1,500 km. Thus, beginning from the second half of the 17th century, Montreal became a point of arrival. Very soon afterwards, it also became a point of departure. Filled with dreams of the Orient and Cathay, the Europeans set forth from Montreal on a route they hoped would lead to the Western sea, and from there to China. Upon arriving in New France, Cavelier de La Salle, following in the footsteps of the likes of Verrazano, Cartier, and Champlain, dreamed of discovering this very route. Later, his seigneury located near the rapids was mockingly dubbed La Chine—the name in French for China.

The obstacle that gives birth to a city must be removed at some point or another—and a canal and bridges eventually occupied a portion of the landscape. At one time, the voyageurs of the North West Company set out from Lachine in search of furs; later, they were followed by the men of William Price Hunt who were headed for the mouth of the Columbia River. But once the canal was built, products of every kind could be shipped to the huge domestic markets of Canada and the United States. The canal also created sources of hydraulic power that prompted major industrial establishments to settle along its banks. Each lock was a prized source of energy during the pre-electricity era.

All manufacturing sectors were to be represented along the canal—from foodstuffs (with Ogilvie) to paints (which were dominated by Sherwin Williams), not to mention the construction of locomotives and cars, which in turn triggered the growth of small firms specializing in wood finish work, upholstery, and wadding. From leather making to iron and steel production, the potential for spin-off was enormous, and specialization became a prime avenue of development. For, to take one example, the Lachine Canal sector certainly produced textiles and garments, but it also turned out wrapping for telephone wire and cloth for covering passenger car seats. Energy was to hand, the prospects for production were good, and services were quick to spring up in the vicinity.

Large industries attracted other large industries, which in turn prompted smaller companies to settle nearby. All told, more than 600, and possibly up to 1,000 companies, came into existence along the Lachine Canal between 1840 and 1940. The impetuous rapids were to give birth to industrial development on an unprecedented scale, making the canal the leading industrial sector in Quebec and Canada, at least in its time.

The most amazing aspect of this story consists in how the creation of thousands of jobs locally in no way prevented the mass exodus of French Canadians. By tens of thousands, they left for New England even as factories were constantly opening in and around Montreal. What factors accounted for this development? Some might look to a problem of language, as the majority of area factory owners were English-speakers, and for many employees the language of work was English. But the answer may not be so simple. The Lachine Canal corridor was home to several examples of French-speaking industrialists, shipowners, and financiers with business connections or who even created financial, industrial, or shipping empires.

One of the initial developers of the Lachine Canal, Joseph Masson, took an interest in lucrative sectors such as coal gas and served as vice-president of the Bank of Montreal, thus becoming one of the leading French-speaking businessmen of the 1830s and 1840s. As it so happened, the Bank of Montreal supported the efforts of numerous industrialists established along the banks of the canal. His example was to be followed, in the financial sector, by Louis-Joseph Forget and his nephew Joseph-David Rodolphe Forget; not only did both men hold major interests in gas companies, they also sat on the boards of directors of several companies whose factories were located along the canal.

The Forgets also had ties to shipping companies and shipowners through the Richelieu Company, which became a component of the omnipresent Canada Steamship Lines in 1913. The Richelieu was founded by a group of French-speaking shareholders that included Jacques-Félix Sincennes, who stood out for his entrepreneurship and business acumen. Sincennes had a portion of his fleet built by Augustin Cantin, one of the leading shipbuilders in Montreal and undoubtedly the biggest such operator along the Lachine Canal. Other examples can be given of what amounted to a French-speaking network of entrepreneurs.

Although the jobs created in Montreal were unable to offset the exodus, they did enable the city to recover its majority French-speaking profile as of 1871. In short, the canal contributed to the dynamism of Montreal by attracting a portion of the population emigrating from the countryside as well as immigrants of all origins. In 1911, 11% of Montreal's inhabitants were of neither French nor British stock.

Besides language, the training and preparation of the workforce was an additional factor. Were the Irish, Italian, and German immigrants better prepared than the French Canadian farmers newly arrived from the countryside? A part of the response surely depends on a breakdown of employees according to their origins. But any definitive answer will have to wait until company archives are opened to researchers and access to information takes precedence over the protection of private information.

This work of research on the Lachine Canal opens onto other avenues of inquiry, particularly in connection with the important role played by energy. On this score, too, the Forgets occupied centre stage as shareholders of the Montreal Light, Heat and Power Company, which held a monopoly over the distribution of electricity on Montreal Island. This company sought profit before all else. By way of comparison, in the early 20th century, the province of Ontario decided to nationalize electricity for the purpose of making this energy source into a lever of economic development. Montreal Light, Heat and Power, on the other hand, continued to charge inflated rates until it was finally nationalized in 1944. For want of any conclusive answer at this time, the question remains as to what extent factories looking for cheap power were discouraged from setting up in Montreal because of how long it took the provincial government to step in.

Successive phases of industrial development have deposited layer upon layer of remains that bear witness to a century of activity along the Lachine Canal. The history of the canal contains elements that are essential to understanding the history of Québec. This collection of documentary photographs will enable readers to cast an attentive glance at a neglected chapter of our history. How can one hope to study the history of Québec, and above all that of Montreal, without carefully examining the story of the takeoff and development of industry born of the turbulent waters of the Lachine rapids? Sooner or later, the impact of canal-building along the St. Lawrence on the economic development of Montreal and the surrounding area will have to be studied.

The Publisher

Introduction

> Mills, grain elevators, warehouses have sprung up in solid blocks in front of the wooden houses, robbing them of the breezes from the country, stifling them slowly. The houses are still there with their wrought-iron balconies and quiet façades. Sometimes music penetrates the closed shutters, breaking the silence like a voice from another era. They are lost islands to which the winds bear messages from all the continents, for the night is never too cold to carry over alien scents from the warehouses: smells of ground corn, cereals, rancid oil, molasses, peanuts, wheat dust and resinous pine.
>
> [...]
>
> In the spring, to be sure, the nights ceased to be quiet. As soon as the channel was free of ice the sirens blow from sunset to dawn, echoing from the bottom of St. Ambroise Street over the entire suburb, and even as far as Mont-Royal when the wind blew that way.
>
> The house where Jean lived was opposite the drawbridge at the corner of St. Augustin Street. Flatboats passed before his door, tankers reeking of oil or gasoline, lumber barges, coalers, all hailing him with three powerful blasts of their sirens—a salute in passing and a cry for liberty, for the open sea they would reach some day when they had left the cities behind and dipped their keels into the waters of the Great Lakes.
>
> Gabrielle Roy, *The Tin Flute* [Translation of *Bonheur d'occasion*, by Hannah Josephson, Montreal: McClelland and Stewart Limited (New Canadian Library no. 5), 1947; reprinted in 1964]

In her novel *The Tin Flute*, Gabrielle Roy successfully encapsulated the reality of the working-class neighbourhood of St. Henri through a wealth of astute, minutely observed descriptions. This same acuity of perception is apparent in the short excerpt above, which highlights several of the dominant features of the Lachine Canal landscape of 1939. The Dominion Textile cotton mills (Merchants Manufacturing and Colonial Bleaching and Printing), the elevators (belonging to Dominion Flour Mills or Canada Malting), and the warehouses had, by then, been rooted in the soil of the neighbourhood for several decades, and many of its inhabitants also worked in the sector. On the eve of World War II, St. Henri was still home to several of its original wood structures, vestiges of the long-ago 19th century, although brick duplexes and triplexes had been cropping up for some time already. In the book, the small furnished apartment Jean rented in the house is a direct allusion to tenant culture, a phenomenon typifying Quebec in this era.

The novelist also alludes to the reality of ship traffic, mentioning the flatboats, coal steamers, and oil ships that sounded their sirens when they drew near the Canadian National railway bridge at St. Augustin Street. The freight borne by these ships exuded a variety of odours which mixed with the smells wafting down from the grain elevators and the warehouses. From that point in the story, however, the surrounding sights, smells, and sounds begin to subside. The urban environment recedes into the background. With the looming and then retreating presence of ships nearby, reality blurs into the realm of the imaginary, as though the mind could light out in response to the call of the open sea. Reverie takes over; it is, perhaps, the only escape route open to human beings mired in their daily existence.

Gabrielle Roy caught—albeit partially—life along the Lachine Canal and its surroundings. She was fully aware of the bleak and brutish aspects of the canal sector; it was literally walled in by a practically unbroken line of factories and warehouses that offered their occupants access to the canal and its transshipment facilities. Concerning the maritime aspects of the canal, however, Roy's grasp tends to be superficial: from her perspective, St. Henri was the world of factories and workers. In her account, St. Henri is merely traversed by this artificial infrastructure from which few of the area residents derived a living. The canal was only a supporting actor. The story's protagonists and the Lachine Canal structures took paths that failed to intersect, with little to span their separately evolving existences. And yet, everything appearing in this passage, whether in connection with St. Henri or any of the wards and municipalities lining the canal, is interconnected. For exploring just these convolutions, the historical study now takes over where the novel leaves off.

From this point of view, the descriptions contained in *The Tin Flute* have to be mined for the content lying just beneath the surface. The industrial wall is indeed a fixture of the environment, giving the canal a structuring role in the urban development of the surrounding wards and municipalities. Industries sought not only transshipment facilities but also phenomenal quantities of water for a variety of needs. In the case of Colonial Bleaching, this water was used to bleach raw cotton and supply boilers. However, the same factory employed a mostly female workforce, which in some cases, was paid ridiculously low wages. In this context, how was a worker supposed to save up enough money to buy a parcel of land or a house? In the best of circumstances, workers could perhaps afford a furnished apartment; failing this, they would have to make do with a wood house that was poorly insulated, heated, and lit—a breeding ground for disease. In such an environment, the radio blaring out its feverish, frantic music curiously opens up a new avenue of escapism.

Merchants Manufacturing did not, however, receive raw cotton, its raw material, directly by boat. The reason why is that the ocean-going ships bearing this precious cargo were unable to travel the entire length of the canal. This freight had to be transshipped at the entrance to the canal at reach no. 2. From these docks, cotton was carried either by train or truck to the mill. Nonetheless, there was an element linking canal, ship, and factory—it was water, once again. For the numerous firms making up the industrial wall along the canal all required water either for their energy needs or for supplying their industrial processes; this water was pumped directly out of the canal. The canal system was dotted with locks, which delimited bodies of water known as reaches. Reach length and the factories' water requirements were two variables that engineers had to factor into any scenario devised to ensure that water levels in the reaches were high enough to satisfy industries and canal-bound ships alike.

And there is still more to the story, for the Lachine Canal stands out like a scar across the land of southwest Montreal Island, the result of a miles-long incision made in order to circumvent a serious obstacle to navigation. It is important to view and understand this dimension of the canal. We have—for reasons of space but by no means out of disinterest in the subject—refrained from commenting on the environmental modifications that resulted from the digging of the canal. The geographic hindrance represented by the Lachine rapids made Montreal a nodal point of marine shipping—i.e., what is termed a breaking point in bulk transport. In Montreal, there was no choice but to transship cargo and change boats in order to continue travelling further inland. Against this backdrop, the building of the Lachine Canal stands out as one part of a greater whole; it lay at the centre of the development of southwest Montreal Island and the complete transformation of the surrounding landscape.

In terms of physical environment, the transshipment of vessels heading either upriver or downriver obviously accelerated the erection of warehouses, docks, and other infrastructures, such as marine towers or unloading cranes. In keeping with the requirements of operating the canal, huts had to be built for lock masters, and bridges constructed for nearby residents. Canal specifications also led to the creation of new types of boats adapted to lock contours. Regionally speaking, however, the evolution of the canal spelled the end of the village of Lachine's reign as the main inland port of Canada. From the time of the French Régime, it had been the jumping off point of the fur trade; henceforth, transshipment would be performed in Montreal.

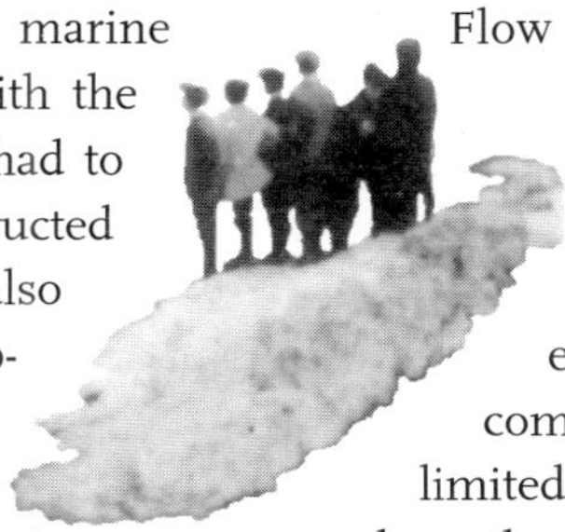

Thus, the Lachine Canal became the head of an inland navigable waterway, the longest such system in the world. Furthermore, this far-flung shipping route confirmed the stature of Canada's merchant marine as a viable enterprise; to be sure, it was a fledgling affair in comparison with the American fleet, but it nevertheless constituted another, if little-known, factor in the development of a nascent Canada. Not only did the canal offer docks and warehouses at which these ships could unload their cargo, it also became a meeting point between ocean-going vessels and ships travelling inland, as was demonstrated by the layout of infrastructures.

This waterway attracted numerous industries seeking cheap energy and water, with some of the largest factories on Montreal Island eventually settling along the banks of the canal. (In passing, "energy" is used here to mean the direct hydraulic power that was developed at specific points along the canal, and should not be confused with hydroelectricity, which is generated by large dams.) The canal system was built around a series of locks that compensated for changes in ground level. For all intents and purposes, the locks of the Lachine Canal functioned as levers, enabling ships to move from one level on to the next. And, as they were built where the ground fell away abruptly, they also represented prime locations for producing hydraulic power. Flow accelerated as water coursed downhill, thus creating greater force with which to turn waterwheels or turbines. The system worked so well that the canal area was nicknamed "Little Lowell," in reference to the famous textile manufacturing complex in Massachusetts. However, only a limited number of such locations were available along the canal. Additional steps would have to be taken in order to find another energy source for the ever-growing number of industrialists who wanted to settle along its shores.

Steam and its faithful companion coal were to step into this new role. Once again, the full importance of the Lachine Canal as a nexus becomes apparent with the advent of coal shipments from Nova Scotia, Pennsylvania, and even Great Britain. What is more, by siting its shops and switching yards between the St. Lawrence and the canal, the major industrial actor known as the Grand Trunk Railway drove the demand for coal. The canal's position as hub was strengthened accordingly, for not only was the canal the meeting point of ships, it also became a point of convergence for rail transportation. As concerns energy in particular, Montreal industrialists were thus able to bargain with various suppliers in order to keep coal prices low. The one-time "Little Lowell" now became "Smoky Valley." However, World War I rather quickly put an end to the soot-filled skies over Montreal, as coal, which had been requisitioned for the war effort, became too costly for the area's industries.

The advent of electricity, which was not generated by canal water (although small power generating plants had been set up along the canal) but by hydroelectric plants a considerable distance from Montreal Island, proved to be a crucial alternative. On the other hand, the Lachine Canal's factories only converted to hydroelectricity some time after their counterparts in southern Ontario. Firms along the canal turned to this energy source only because the economic times forced them to; for them, this changeover meant relinquishing the precious energy self-sufficiency they had enjoyed since settling along the canal and buying it instead from Montreal Light, Heat and Power, which continually charged higher rates than those in force in southern Ontario.

Be this as it may, the canal continued to hold a certain attraction for industrialists, on account of its facilities for transshipping raw materials and the supply of water they could pump into their factories. For not only was the canal used as a shipping lane and energy source, it was also a supplier of water to industries. Owing to this threefold use, between 1846 and 1945, the canal attracted more than 600 different firms covering the entire gamut of industrial manufacturing, whether light or heavy. In fact, from 1846 to 1945, factories representing each of the 17 major industrial sectors were established along the edges of the canal at one time or another, with some companies eventually taking over from others. Some of these industrial concerns grew to national, continental, or Commonwealth-wide prominence. In the process, the canal corridor became the epicentre of Canada's manufacturing sector, and Montreal the nation's leading metropolis.

The presence of manufacturing actors of this magnitude in turn attracted suppliers to locate in the vicinity. Thus, for example, the establishment of the Grand Trunk shops in nearby Point St. Charles prompted the arrival of sawmills and industries that manufactured wheels, suspensions, and cars for trains, as well as carpentry shops for cabinet and decorative work on passenger cars. These manufacturing concerns in turn generated a demand for the wood needed to build seats and the padding and fabric required to cover them. And this is but one, classic example of a pole of industrial development intertwining with the evolution of the canal landscape as a whole. Also worth noting is the carbochemistry sector and several of its links to the textile industry, or those arising between the manufacturers of textiles and electrical wiring. In short, the critical mass of factories located on the edges of the canal served to catalyze the expansion of regional industry to awesome proportions.

In addition, some of the larger companies set up factories with a workforce numbering in the hundreds if not the thousands. From close to 4,000 workers in 1871, the factories of the Lachine Canal quite probably employed 28,000 during the between-war period—in other words, they accounted for more than one out of every five manufacturing jobs on Montreal Island. This phenomenon produced two main types of impact. First, it prompted these workers and their families to live in the vicinity of their workplace, thus adding considerable impetus to urbanization on both sides of the canal system. Second, while not necessarily belonging to union organizations, these workers assembled within their respective factories to claim better conditions in respect of workplace hygiene, working hours, wages, etc.

The use of the term "workman" should not be construed as obscuring the presence of women in factories or restricting their participation in the manufacturing workforce to times of war. Women had been working in light industry such as cotton mills since the 19th century. The 20th century saw them take jobs in other industrial sectors such as electrical apparatus and supplies. This gender-based shift in occupational profiles, coupled with the emergence of unions (which albeit occurred unevenly in the canal area), provides evidence of a social transformation.

If workers and their families decided to live close to their place of employment, this was because their wages were rarely sufficient to cover the cost of transportation using the embryonic mass transit system available during the 19th century. Only in the 1890s was there a full-fledged network of electrified tramways capable of providing service throughout the various municipalities that had formed subsequent to the arrival of these workers. Along the stretch of the canal located nearest to the city of Montreal, the sector was witness to the founding of St. Cunégonde and St. Henri on the north side, and St. Gabriel and Côte St. Paul on the south side. The St. Anne ward, much of its population consisting of the Irish immigrants who had dug the first canal, had already been in existence for some time. At the west end of the canal, the municipality of Lachine began to develop in full, following a decades-long lull, when it granted certain financial concessions with the objective of attracting industries within its legal boundaries. Between both of these inhabited concentrations, Ville Saint-Pierre and Ville LaSalle timidly evolved into full-fledged municipalities.

It is thus apparent that economic imperatives largely drove the birth and growth of these new municipalities. Their development was closely bound up with the fortunes of real estate speculators, who accorded little room for such things as beautification or public spaces. Similarly, since urban development was to a great extent the work of developers, considerations of profit dominated the use of the natural and the built environment. Thus, housing conditions sometimes left something to be desired, particularly in the case of households that were relegated to "back courts." The canal area was a most infertile ground for any would-be garden city, and was all the less likely to become so during the steam era, when soot literally showered down upon the area. This air-borne pollution, once deposited, formed layers of filth worthy of an archaeologist's trowel and was inevitably a source of numerous health problems for workers living in the vicinity of the dense industrial complex. As working-class families were already beset with numerous public health problems such as infant mortality and tuberculosis, living in such a grey environment could only complicate health matters further.

The insalubriousness of housing was an issue that had its counterpart in the factory environment, which was often poorly lit and above all poorly ventilated. Without a doubt, the lack of proper ventilation contributed to the spread of infectious diseases. However, workers also had to beware of the machines assigned to them, since a single moment of inattention could quickly cost them several days' wages. These wages, even when full, were inadequate to afford decent housing, clothing, or food. If by mischance these men missed any workdays, their wives or children would have to be called up for active duty in the employment market so that the family could make ends meet.

Although legislation stipulated that only children over 12 were authorized to work (subsequently the minimum legal age was raised to 13, then 14, and finally 16), there is reason to doubt that the law was seriously enforced. Labour legislation as a whole echoed developments in the factories along the Lachine Canal, for legal measures represented a response to labour protest movements centring on shorter working days or improved wages. All in all, the history of the Lachine Canal is also the history of Canadian and Quebec labour relations.

It is also the history of visionary businessmen, who succeeded in making the right decisions at the right time. Those who were slow to act or who made the wrong decision were eliminated. It is also the history of shopwork, as experienced by our grandfathers and great-grandfathers. With its weirs and sluices, the Lachine Canal attracted industries and workers to what has been termed the cradle of Canadian industrialization. It is important to realize that the canal, in addition to being a shipping lane, also represents an industrial corridor, and as corollary to this, an extension of

Montreal's urbanization as this process unfolded during the 19th and 20th centuries. While occasionally opposed to one another, all three dynamics—shipping, industrialization—and urbanization—were nonetheless interdependent. With its 150 photos and accompanying texts, this book has been designed to provide the reader with an overview, which, as we are only too aware, uses only the broadest lines to sketch out the history of the Lachine Canal.

Yvon Desloges
Alain Gelly

The Era of Canals

Bunnett, H., *Old Lock, Lachine Canal*,
National Archives of Canada, C 113716

FOR A FULL CENTURY, between 1760 and 1860, the Western world was witness to a canal craze, which first took hold in Great Britain and then spread to the Continent before crossing the Atlantic. Twenty-three American states, particularly those bordering the Great Lakes, launched into this adventure; in Canada, canal building was undertaken primarily with the objective of facilitating communications and trade between the St. Lawrence and the Great Lakes.

The British "Canal Mania"

In 1761, when the third Duke of Bridgewater inaugurated his canal in Worsley, built according to plans by Thomas Telford, it attracted the interest of manufacturers who saw this as a means for carrying British coal more easily and at lower cost. However, the idea of canalizing rivers throughout Great Britain was nothing new; as early as 1708, Parliament, had passed the first legislation for this purpose, a bill to make the Kennet River navigable. During the following half century, 1,600 kilometres of rivers were made open to navigation. However, these rivers did not make it any easier to move coal, and road transportation, which was time-consuming and costly, still had to be relied on.

The Worsley Canal showed the value of water transportation for hauling coal from collieries to the industrialized regions. Its success created a vogue for this mode of transportation that was to continue for nearly 80 years. The golden age of British canals reached its peak in the 1790s, when the British Parliament authorized the construction of 55 canals. These local and regional infrastructures, the work of small contractors, consisted in narrow waterways managed by local companies; this development would nevertheless give birth to a profession, that of civil engineer. In Great Britain, only a few canals were designed for ocean-going vessels. Around 1840, the waterway system, which covered some 6,500 kilometres, served most of Britain's cities and industrial districts.

However, by making such quantities of coal readily available, canals opened the way to their eventual undoing, as railways rose irresistibly to prominence. Wearied of monopolies and canals of unequal width, industrialists very quickly subscribed to the notion that railways could compete with waterways and thus intensify competition. As of 1830, the canal linking Liverpool to Manchester was faced with direct competition from a railway line. The interest in railways surged, with 170 authorizations for rail companies being granted in the early 1840s. Waterways and railways then began competing fiercely with one another, which had the effect of bringing down freight charges. In 1850, Great Britain had more than 9,600 kilometres of railways, and canals began falling into the hands of rail companies or went bankrupt. In the long term, only industrial canals managed to survive.

The American "Canal Era"

The United States followed closely upon the British lead. The country's civil engineers, who drew heavily on developments in Great Britain, began building the first canals during the 1790s, at the peak of the British canal mania. However, it was only after the War of 1812, and with the building of the Erie Canal in particular, that the canal era genuinely commenced. This period would end with the U.S. Civil War. The Erie Canal, begun in 1817 with the admitted goal of diverting trade from the West toward the Hudson River and New York City, was completed in 1825. Considered to be the most outstanding American financial and technological accomplishment of its time, the Erie Canal underwent a widening project that began in 1835 and continued until 1862. When this project was completed, the canal measured 21 metres wide and 2 metres deep along its entire length.

All the states bordering the Great Lakes were swept up in the American canal craze—not to mention Virginia, whose canals, open to traffic all year round, were intended to siphon trade away from the Ohio River valley and Pennsylvania to the Chesapeake Bay, at the expense of New York City. However, in the United States, as elsewhere, rail transportation began to predominate during the 1850s.

The U.S. canal craze began to subside in 1836, not long after it had been decided to broaden the Erie Canal and cross through the mountains of Pennsylvania. The following year, a financial crisis triggered by the bankruptcy of the Morris Canal and Banking Company of New Jersey tempered enthusiasm. Nevertheless, around 1860, the system of American canals covered 6,800 kilometres throughout some 23 states. The Erie Canal, as well as several other canals such as the Chesapeake and Ohio, held up against the pressures of railway development, but the tonnage carried by these waterways plateaued as of 1872. American canals, which were designed to promote trade and commerce between the inland areas and the markets of the Atlantic seaboard and which were built by private enterprise, came under the jurisdiction of the states that funded or underwrote them, not under that of the federal government.

The St. Lawrence Canals

In Canada, the situation differed considerably from that in the United States, since the construction of canals was funded and supervised by the imperial government, if not by the colonial government or the federal government. At the time of Confederation, Canada had three main canal systems—the St. Lawrence system, from Montreal to the Great Lakes; the Ottawa River-Rideau Canal, an inland shipping route between Lake of Two Mountains and Kingston (via Bytown, as Ottawa was then called); and the Richelieu River system from the St. Lawrence River to Lake Champlain.

Along the St. Lawrence, the earliest system served a military purpose. From 1779 to 1783, during the American Revolutionary War, the Royal Engineers supervised the construction of locks between Lake St. Louis and Lake St. Francis. They thus sought to improve military communications between Upper and Lower Canada, although commercial use was also anticipated. Between 1804 and 1817, this first canal system underwent substantial modifications on account of three factors: the massive arrival in Loyalist immigrants from the United States and their subsequent settlement of Upper Canada; the War of 1812 between Canada and the

United States; and the advent of the Durham boat, a vessel much wider than the "batteau"[1] in use until that time. As a result, the dimensions of the military canals were doubled, to 3.6 metres in width and 1 metre in depth. However, cargo still had to be transshipped and transferred overland from Montreal to Lachine to bypass the rapids on the south shore of Montreal Island.

Assured of a contribution by the British Empire in return for unrestricted access for its military ships, the government of Lower Canada launched into building a canal in the southwest sector of Montreal Island between Lachine and Montreal in 1821. It took no less than four years to complete the Lachine Canal. Measuring 15.5 kilometres long, this waterway surmounted a difference in ground level of 13 m through a corridor 15 metres wide. The first canal included seven locks measuring 30 metres by 6 metres and 1.5 metres deep. During the 1830s, the government of Upper Canada single-handedly undertook construction of the Cornwall Canal because the population along the shores of the St. Lawrence feared its trade activities would decline once the Ottawa-Rideau system was brought into service. However, it took several years before the work was actually completed. Respecting the Great Lakes system, it is also worth pointing out that the Welland Canal, between Lake Erie and Lake Ontario, was the work of private enterprise, a portion of whose capital originated in the United States. Begun in 1824 and completed 10 years later, this project also benefited from a colonial government loan.

In 1841, on the eve of the Union of the Canadas, the St. Lawrence system presented the same shortcomings as the British and American canals: it was an incomplete network of different dimensions, as the military canals were considerably narrower than the Lachine Canal. Hoping to take advantage of a favourable economic situation and grab a share of the exports originating in the American Midwest, the government of the Union of the Canadas, with the backing of an imperial guaranteed loan, began work standardizing the St. Lawrence canals. In the 1840s, the Lachine Canal was widened; the smaller military canals were replaced by the Beauharnois Canal; the Cornwall Canal was completed; the Williamsburg Canals upriver from Long Sault were developed; and the Welland Canal was widened. From that time on, Great Lakes ships could ply the waters as far as Montreal. The locks measured a minimum of 61 by 14 metres, and were 2.75 metres deep.

Although canals in England and the United States were being supplanted by rail transportation, the then nascent Canada assembled a board of inquiry on the future of canals in 1870. Not only did the Canadian government accept the recommendations of the board to upgrade the canal system, it also made this project a central theme of national development: improvements to inland navigation were essential to the development of interprovincial trade. The adoption of this policy was the outcome of three immediate factors: the Confederation of 1867 and an image then forming of a new country to be developed, the abolition of the Reciprocity Treaty with the United States in 1866, and the rise of interprovincial trade during the previous few years. In fact, the political authorities sought to create a domestic market in which coal from Nova Scotia could be shipped as far as Toronto and farm products from Ontario and the new territories in the West could find outlets either in Quebec or on international markets.

As a result, the St. Lawrence system was improved and locks were standardized at 82 metres in length by 14 metres in width (hence the same width as before). The standard depth of canals was eventually increased to 4.25 metres, in reaction to pressure by shipowners. Since navigation had to continue unabated, work

1. Batteau: a flat-bottomed river boat, fitted out with sails as needed and capable of carrying up to 4 tons of goods.

on canal standardization was pursued until 1904. The project had been launched in Montreal in 1874, with the third Lachine Canal being completed in 1883. It targeted much more than the St. Lawrence system alone, for it also entailed performing work on the Welland Canal, developing new canals in the Great Lakes area and the Maritimes, and improving existing canals in the Ottawa and Richelieu River valleys. In the St. Lawrence sector, the Beauharnois Canal was even replaced by the Soulanges Canal as part of work completed between 1892 and 1899. The Sault Ste. Marie lock, built from 1892 to 1894, enabled some ships to travel from Lake Superior as far as Montreal.

Although the St. Lawrence system was in great measure the product of the canal craze in Great Britain and the United States, the size of locks in the St. Lawrence system nevertheless reflected a quite different state of affairs. British and American canals were narrow, since they were primarily inland canals for barge traffic, whereas the wider St. Lawrence canals were intended for the movement of sailing ships or steamships. As well, some were industrial canals—that is, the water circulating in them was also used for industrial purposes, whether to meet energy needs or to supply water for industrial processes.

The Lachine Canal and the Surrounding Industrial Complex

In an era when navigation represented the most efficient means for carrying people and goods, Montreal represented a "breaking point" in navigation toward the interior of the continent—in other words, boats had to be unloaded at this location in order to continue sending goods on. Building canals along the southwest point of the island represented a project dear to the Montreal administrators of the French Régime. By the late 17th century, work had been started on bypassing the Lachine rapids—an ambitious project that was nevertheless abandoned after several years of work. More than a century would go by before the first ship could make its way through a navigable corridor and thus skirt the rapids.

In 1825, the government opened the entire length of the Lachine Canal to navigation. The canal became a victim of its success, as marine traffic increased enormously. The number of ships passing through it annually increased sevenfold between 1825 and 1841, while the number of passengers heading for the Great Lakes increased fivefold, in keeping with a steadily growing migratory movement. A geographic barrier had turned into an economic advantage.

Gradually, the import-export sector became the cornerstone of the Montreal economy, as the acknowledged goal of shipowners was to make the city a hub for transiting ships and products from western Canada such as lumber and grain, which depended on Empire-wide demand and preferential tariffs. However, the ambitions of this project were frustrated by the creation of the Erie Canal, which had the effect of generating brisk competition between Montreal and New York City merchants for control over trade from the interior of the continent. Pursuing this commercial objective entailed constructing warehouses and other storage facilities and employing an abundant workforce for handling and transporting goods. The increase in activity attracted new inhabitants, and this in turn stimulated local production. It was against this backdrop that Griffintown, the core of the future St. Anne ward located at the Montreal end of the canal, developed between 1815 and 1830.

Increasing ship traffic and ship dimensions made it necessary to widen the Lachine Canal and the other St. Lawrence canals. This project would provide steamships access to the Great Lakes, reduce transshipment between Kingston and Montreal, accelerate navigation, and lower transportation charges. Work on widening the canal began in 1843. Five years later, the canal had become 36 metres wide (instead of 15 metres) and the number of locks had dropped from seven to five, including the two in the vicinity of Montreal harbour, where the difference in ground levels was more than 7 metres. The result was shortened lockage time and accelerated navigation, both improvements that further reflected the commercial concerns underlying the project.

Widening the canal also increased flow, and when this strong flow was combined with a considerable drop in water level (as was the case with the locks of Montreal harbour), enough energy was created for supplying a mill's needs. Whereas in 1851 only a dozen steam engines were in operation to supply the machines of Montreal factories, the opening of the new waterway created a particularly attractive alternative in the form of hydraulic power.

The underlying principle was already known and used in numerous rural establishments (sawmills, gristmills, and others) and had, as well, been implemented in a number of American urban centres. Now, it only remained to be applied to the canal. The Public Works Commissioners decided to lease hydraulic power in order to increase canal revenues. Three sectors presented a good potential for energy: canal reach and lock no. 2, in the harbour sector, with 7 metres of difference in level; the St. Gabriel sector, which presented close to 3 metres in level change; and the Côte St. Paul sector, also with close to 3 metres in level change, although its location at a considerable remove from urbanized areas constituted a serious handicap. Montreal entered a period of water-powered industrialization when, on November 23, 1846, the government put up for lease the first hydraulic lots, measuring 24 metres along the canal front located in reach no. 2—that is, in the vicinity of today's Mill Street. So doing, it did away with one of the main obstacles to industrial development in Montreal—the lack of suitable energy sources. At that time, steam-powered machinery was expensive, required considerable upkeep, and demanded qualified operators as well as great quantities of fuel.

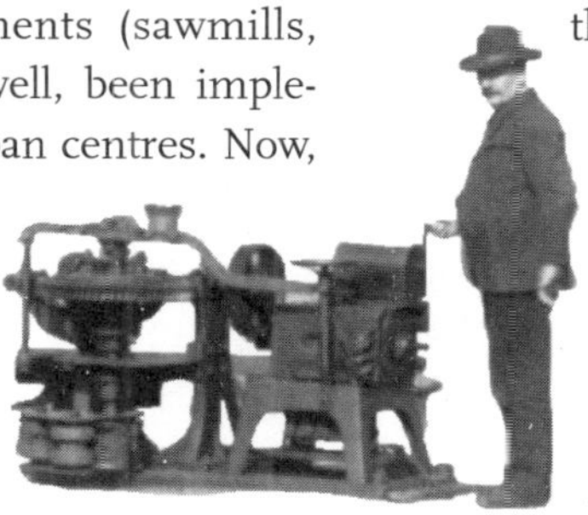

Although not all of the Mill Street hydraulic lots found takers, the Public Works Commissioners nevertheless decided, beginning in 1850, to lease all lands suitable for the use of hydraulic power in the St. Gabriel locks sector. This time, the government did not take part in the project; instead, it drew up leases with businessmen who in turn sublet the various hydraulic lots after first having established infrastructures on them. Three years later, William Parkyn leased the water privileges of Côte St. Paul. Handicapped by his distance from the urban centre, Parkyn had to build not only the headraces but also the industrial establishments in this sector, which soon proved his undoing. Nonetheless, his hydraulic complex would survive into the 1960s.

Water-powered industrialization constituted one of the three factors behind the industrial and urban boom enjoyed by southwest Montreal. The others were the construction of the Grand Trunk Railway infrastructures and the obligation imposed on the Sulpicians of divesting themselves of their St. Gabriel farm and subdividing it into lots. With the gradual establishment of industries along the canal shores, workers the began to seek out affordable areas in which to settle. In the absence of means of communication and mass urban transportation during this period, they had to congregate near their places of employment.

With the creation of hydraulic lots, the canal took on a triple function: it served both as a waterway and as an industrial water supply system offering a source of energy and water for use in manufacturing processes. These mutually opposed functions led to a conflict between shipowners and industrialists. Industrialists needed steady flow and rapid current; shipowners, on the other hand, required a slow current. It was a conflict that would be aggravated by the industries' unquenchable thirst for water with which to satisfy the ever-increasing demands of their facilities. As a result, engineers attempted to reconcile these two irreconcilable interests through piecemeal solutions—either by developing weirs, widening the canal itself in some sectors, or even increasing the water supply at the head of the canal.

A Bird's-Eye View

The Basic Principles and Infrastructures of Canals

Lachine Rapids, Bibliothèque nationale du Québec, E.-Z. Massicotte albums, undated

SKIRTING THE UNAVOIDABLE

When Jacques Cartier sailed up the St. Lawrence in search of the riches of Cathay (China), he encountered a formidable obstacle to the south of Montreal Island: a daunting series of rapids that descended 13 metres over some 14 kilometres. No subsequent French or British navigator would be able to surmount these "impetuous rapids," at the base of which the current flowed at more than 8,500 cubic metres per second. Boats and timber rafts were able to descend, or "shoot," the rapids unimpeded. However, as boats were unable to travel upriver, their cargo had to be unloaded in Montreal and carried to the western tip of the island—an operation that could take up to two days. From there, the voyageurs and coureurs de bois (that is, French fur traders or woodsmen—literally, "runners of the woods") set out on their expeditions in quest of furs. During the 18th century and at the turn of the 19th century, the village bearing the evocative name of Lachine (from the French name for China) was Canada's leading inland port and the jumping-off point of the fur trade's transportation system. Although the fur trade began to decline toward the early 19th century, the Hudson Bay Company maintained a warehouse at Lachine until mid-century.

Old Lock Number Two, Looking East, National Archives of Canada, Department of Public Works collection, PA 110242, 1908

THE PRINCIPLE OF COMMUNICATING VESSELS

At each abrupt drop of the land, a lock was built. The harbour sector had two locks, as the change in water level between the canal and the harbour varied between 7 and 8 metres, depending on the level of water during the late summer or in spring. A lock, a mechanism for lifting or lowering boats, works according to a relatively simple principle in which water is used as a lever. A boat descending with the current (in the case of this canal, from Lachine to Montreal) entered a lock *chamber** that had been previously filled to the same level as the *reach** through which the vessel had just passed. Once the gate (i.e., the head gate) was closed behind it, the water contained in the chamber flowed out through the sluice valves located along the bottom of the gate and into the reach below. When the boat had been lowered to the same level as the water in the adjacent reach, the gates only had to be opened so that it could continue on its way. Conversely, a boat travelling upriver first entered the emptied lock chamber from the lower reach. Once the gate was closed behind it, the boat rose as the water of the upper reach filled the chamber. As soon as the water levels of the chamber and the upper reach were equal, the gate was simply opened and the boat resumed its journey. It is worth pointing out that the 1825 version of the canal had three locks in the harbour sector, but was given only two under the canal layout of 1843-1848, thus enabling shipowners and navigators to save one lockage procedure.

* *chamber*: the portion of a lock located between two gates.

* *reach*: a body of water stretching between two locks. The upper reach lay upstream from the lock it flowed into while the lower reach lay downstream from the lock emptying into it.

Lachine Canal, Lower Gates, National Archives of Canada, Department of Public Works collection, PA 118792, 1904

MAKING A GRAND ENTRANCE

Since a lock implies a change in level, the head gate is where the sill—or, the declivity—is located. Once the head gate is closed, it acts as a dam, retaining the water of the entire reach. The gates are opened only when the level of water in the chamber is the same as that of the reach, thus cancelling out the pressure of the inflowing current. Obviously, if the head gate were to open inwards when the lock was empty, the pressure of water would quickly damage the sides of the lock (called "lock walls") and the floor (called the "lock bottom"), as well as the tail gate. In order to hold back such a considerable mass of water, gates had to be exceedingly solid. Built from wood, they weighed several tons. The lower portion of each gate was fitted out with sluice valves, which were operated by means of cranks located on top of the gate. One detail worth noting is that the lock gates were always set at an angle to the current. In our climate, this method of installing gates had the advantage of preventing them from accidentally opening as the result of a sudden inrush of water during spring flooding or the break-up of ice. In the event of a collision or operational failure, a gate could be speedily replaced. Canal employees were able to make their way over the canal water with a floating crane in order to move replacement gates, which were stored in the vicinity of the locks. Such readiness was all the more crucial in that ship traffic could not be interrupted for too long.

Tate's Dry Dock, Looking East, National Archives of Canada, Department of Public Works collection, PA 109976, 1916

HYDRAULICS 101

The engineers who supervised the construction of the canal had to factor several basic hydraulic data into their calculations. For one, they had to make sure that the supply of water satisfied lockage needs, compensated for leaks, and offset evaporation during summer. When, as the second canal was completed, hydraulic lots and dry docks were leased out, not only did engineers have to incorporate these new data, they also had to account for the water consumption of various factories. In the Erie Canal, the main competitor of the St. Lawrence system, the quantity of water lost to leakage could amount to 20 centimetres in water level every day. In the case of a canal that was shallow and very narrow to begin with, such a factor could represent a serious handicap for vessels since they could run aground without warning. Major leaks occurred in the Lachine Canal, but no engineer tracked losses on this basis. However negligible evaporation might be in the St. Lawrence system, the phenomenon nevertheless entailed a loss of water in the case of a reach. A locking operation could move up to 3,500,000 litres of water, the equivalent of the contents of 67 above-ground swimming pools measuring 7 metres in diameter, a quantity that obviously has to be updated according to the number of ship passages. A dry dock like Tate's contained close to 13,650,000 litres of water alone, or the equivalent of 250 pools. In addition, engineers had to consider the water that was diverted either into a supply line for producing hydraulic power or directed into factories to meet various manufacturing needs. These movements in water masses represented only the first side of an equation that engineers were responsible for solving.

Lachine Canal from Côte St. Paul, Looking East, Parks Canada, Quebec Canals, undated

THE OTHER SIDE OF THE EQUATION: THE TANGIBLE VS. THE INVISIBLE

The second half of the equation facing engineers during the 19th century concerned the notions of flow, slope, and velocity. The leasing of hydraulic lots to industrialists confronted engineers with a dilemma: supplying turbines with water implied accelerating the current along the canal and would be detrimental to navigators. On the other hand, maintaining the velocity of the current at a level acceptable for ships would force industries to resort to an energy source other than water power—but the acknowledged goal of canal widening had been, precisely, to create this power. In the opinion of shipowners, the rate of flow had to be maintained at 1.6 kilometres an hour. This was, however, an ideal figure, for velocity depended on the slope represented by the first reach: the gentler the slope, the more current was favourable to navigators. As the second canal featured a slope of 34 millimetres over a distance of 7 kilometres, the engineers proposed doubling the figure to 76 millimetres. This degree of slope may seem negligible, but the resulting change in ground level had to be adjusted in keeping with the depth of basins at the harbour end downstream and, obviously, with the extent of the drop occurring at locks. The last factor to be considered was the inflow of water into the canal—that is, the rate of flow. When the canal's engineers performed their calculations, they assumed that water consumption by local industries would hold steady. And, based on estimated long-term water levels of the river at Lachine, combined with the depth of the regulating lock, they planned on a stable flow of close to 85 cubic metres per second. Yet, in what became an inescapable conclusion, the width of the canal had to be widened and new weirs created, which would reduce the speed of the current while also continuing to supply industries appropriately. In 1874, when the Canadian government decided to widen the canal a second time, it attempted to reconcile these two irreconcilable imperatives. In 1948, out of approximately 100 factories then in operation, seven of that number alone consumed close to 50 cubic metres of water per second!

Ocean Basins

General View, Montreal Harbour, Looking Eastward,
Centre d'archives d'Hydro-Québec, Shawinigan Water and Power Co. collection (F1), F1 / 700862, 1931

BASINS FOR SEAGOING SHIPS

In the vicinity of Montreal harbour, the first section of the canal, which also went by the name of the Montreal terminal on account of its depth, enabled ocean-going ships to advance this far inland. It included two locks, two reaches, and a turning basin. The first lock served as a point of access for ships arriving from Montreal harbour. There was room in the first reach for a few ships to remain stationary while waiting to commence locking as another ship entered or left the lock. The role of this section of the canal was reflected by the presence of several basins and side docks. The photo shows four of the side docks of the interior harbour: they are Tate's dry docks (so named for the first occupant), before which two ships lie side by side. On the opposite shore on the left hand side are two trapezoidal "flour basins," so named because one of their original functions consisted in receiving flour. These basins recently reappeared in the Montreal downtown landscape after having been filled in during the 1960s. Facing the flour basins is the Wellington basin, where two ships are shown moored side by side. The considerable difference in ground levels occurring between the second reach and the harbour sector made the former an ideal location for generating direct drive water power. With this in mind, beginning in 1846, the canal authorities laid out fifteen hydraulic lots along Mill Street for the purpose of attracting industrialists seeking cheap energy. These lots were laid out where the three ships appear moored in a row.

The Montreal Warehousing Company,
McCord Museum, Notman Photographic
Archives, MP 011 / 243, ca 1875

FLOUR BASINS

During the 19th century, wheat and flour were among the main staples conveyed by ship and earmarked for export. Not surprisingly, two new basins devoted to this type of shipment emerged in the landscape of the second canal (1843-1848). These trapezoidal expanses of water were called flour basins on account of the flour barges seen here waiting to transfer loads of flour on to ocean-going vessels. Measuring 106 metres in length along the longest side and 33 metres in width, these basins had a depth equal to that of the main basin and were thus integrated into the port terminal. Beginning in 1846, government flour warehouses appeared along flour basin docks. Montreal lacked such warehouses, however. That is why, in 1865, John Hall founded Montreal Warehousing on the docks of the no. 4 flour basin. There is no better illustration of the canal's role as an interface between inland movement, warehousing, and transshipment of flour onto ocean-going vessels than the scene depicted in this photo taken circa 1875. The upper portion of the building held grain while the lower portion in stone was used to store flour. Two marine cranes are used to unload barges arriving from inland Canada. A ship is moored at the far end of the pier.

Lachine Canal Weir No. 1, Looking North,
National Archives of Canada, Department
of Transport collection, C 063602, 1907

RUNNING THE SLUICES

Occasionally, the canal had to be emptied, either because there was too much water—as the result of springtime flooding, for example—or because repairs had to be made on a wall or a lock. Some system had to be devised for eliminating overflows. Weirs fulfilled this need, allowing the overflow from one reach to be discharged into another, although that occasionally created turbulence and other inconveniences. A weir was the equivalent of a dam, but it eliminated overflow via sluice valves at the bottom of its structure. In the Montreal harbour sector, two weirs were installed on reach no. 2, which lay 5.5 metres lower than the upper reach of the canal at Lachine, so as to prevent any flooding of the factories lining the basin. One was installed in the vicinity of lock no. 2, with the other being located close to Tate's dry dock. In this photo taken downstream from the first weir, two things are particularly worth noting, first, the slope downward toward the harbour, and second, the rapidity of the current. For sharp-sighted readers, the boat in the background is the steamer *Chateauguay* of the Montreal and Chateauguay Navigation Company.

Carre, William H., *Lachine Canal from Grand Trunk Railway Offices*, Bibliothèque nationale du Québec, E.-Z. Massicotte albums, Streets, no. 2-198, 1898

THE WELLINGTON BASIN

In 1874, the Department of Public Works brought out a report on a project to expand the Montreal terminal. This project was to include a basin 68 metres wide by 381 metres long by 6 metres deep for receiving ocean-going vessels. Its walls were to be of masonry, with the wall top to be in ashlar. Around each of these walls, 11-metre-wide wood wharves were to be built. The contract was granted in 1880, and the Wellington Basin received its first ships as of 1883. The project had been conceived in 1862, when the canal commissioners began planning to build deep-water basins surrounded by docks on which to raise warehouses and elevators and thus accommodate trains arriving from the switching yard of the Grand Trunk railway located less than 500 metres further inland, in Point St. Charles. The photo shows Grand Trunk trains picking up or leaving off a cargo of building timber. The flour elevators were never to be erected, being supplanted instead by coaling cranes during the late 19th century. Thus, the basin was to witness the installation of Montreal Coal Elevating, the wholesaler Andrew Baile, or Rochester and Pittsburgh Coal. Today, the basin has been filled in to a great extent.

Black's Bridge, Lachine Canal, National Archives of Canada, Department of Public Works collection, PA 110030, 1923

HYDRAULIC POWER: WHAT'S GOOD FOR INDUSTRIALISTS IS GOOD FOR A CANAL STRUCTURE, TOO

The appearance of Black's Bridge in the canal landscape coincided with the first phase of expansion (1843-1848). Located close to the hydraulic lots of reach no. 2 and lock no. 2, this bridge, which was a swing or pivot bridge like all the other bridges of the second canal, was originally built out of wood. However, it had one particularity that no other canal bridge would have: the bridge was opened and closed using power conveyed by a water wheel. As is clear from this photo, the pronounced declivity justified purchasing the wheel. At the same time, its installation resulted in the elimination of several bridge operators' jobs. On the other hand, wood bridges were notorious for the amount of upkeep they required. As traffic intensified between both sides of the canal on account of the nearby factories, a new, iron swing bridge, with a deck measuring 47 metres long, was built in 1879 by one of the industrialists along reach no. 2, the foundry operator William Bartley. This scale of construction enabled the bridge to span the width of both new locks. This first metal bridge was replaced 25 years later by the one appearing in the photo, which was constructed by Dominion Bridge, a company located near the canal, but in Lachine—in other words, at the other end of the canal. The wood structure appearing on top of the bridge is the bridge operator's cabin. It contained all the instruments required to pivot the bridge, which was powered by electricity as of the early 20th century.

Electrical Machinery, Lachine Canal, National Archives of Canada, Department of Public Works collection, PA 110134, 1909

THE ELECTRIFICATION OF LOCKAGE OPERATIONS

Although electric lighting had been implemented along the canal from 1887, electricity was used to operate lock gates only beginning in 1909. Prior to that time, the lock masters had used winches and chains to move gates. However, the canal authorities proceeded to electrify lockage operations following the construction in 1904 of a small hydroelectric power station next to the Côte St. Paul lock, located several kilometres upstream. In summer 1905, the older of the two Côte St. Paul locks was targeted for electrification, with this process being subsequently extended to the other nine locks. Electrification made it possible to halve the time and the staff required to open gates—an operation that from then on took less than a minute. Whereas previously, four men had been required to activate winches on both sides of the lock, now only two were needed to operate the electrically-powered gates.

Basins Reserved for Inland Fleets

NOTMAN, William, *Montreal Views from Street Railway Power House Chimney*, McCord Museum, Notman Photographic Archives, MP2942, MP2943, MP2944, 1896

BASINS FOR INLAND FLEETS

The Montreal terminal also included four other basins that now no longer exist. This panorama by William Notman, taken from atop the chimney of the Montreal Street Railway powerhouse, offers fascinating insight into activity in this sector during the late 19th century. The docks of these basins were used by lumber and coal companies for loading or unloading cargo arriving from the interior of the continent or headed there. During the 20th century, all these docks would be built over with warehouses used by several of Canada's best-known inland shipping companies, including the St. Lawrence Gulf Navigation Company, the Calvin Shipping Company, and the Tree Line Company (the latter two firms were subsidiaries of Canada Steamship Lines, which also had its own dock). These warehouses could be up to 230 metres long by 12 or even 30 metres wide, as in the case of the Canada Steamship dock, and were leased either in a single block or subdivided into smaller units. In short, these basins were set aside for inland shipping fleets.

Lachine Canal, Basin No. 4, National Archives of Canada, Department of Public Works collection, PA 110243, 1903

LUMBER BASINS OR COAL BASINS?

The first basin to appear in the canal landscape was built in the vicinity of the St. Gabriel farmstead during the first phase of expansion; from this time, it was associated with storing wood in the form of firewood or building timber. However, the lumber merchants soon began to complain of space limitations, as the centre of the basin was occupied by an island. During the 1870s, this island was turned into a jetty separating St. Gabriel basins no. 1 and 2 as part of improvements made necessary by the expansion of the lumber trade. Only a few years later, in the early 1880s, two additional basins were dug, measuring 38 metres wide (that is, 7.6 metres narrower than the previous two basins) and 229 metres long and 4.5 metres deep. The new wharfage area, covering some 40,000 square metres, was fought over by lumber merchants and coal wholesalers. Lumber merchants were entitled to use basins no. 3 and 4, whereas coal merchants were allowed to store their coal on the docks of basins no. 1 and 2. Not surprisingly, all manner of sawmills, door and window manufacturers, cabinetmakers, and lumber yards set up in the vicinity during the 19th century. These basins were 1.2 metres shallower than the ocean basins—an indication that different fleets were used for each of the two navigable circuits.

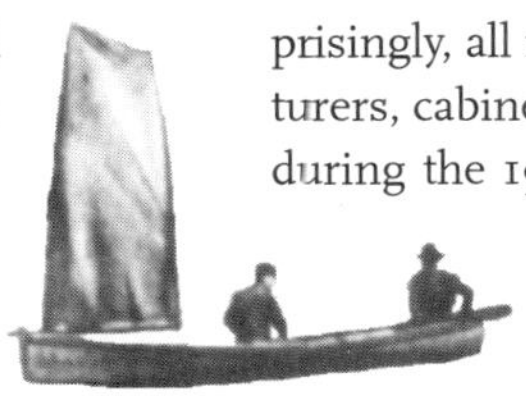

Lachine Canal, Wellington Street Tunnel Excavation,
National Archives of Canada, Department of Public Works collection, PA 119388, 1932.

THE LACHINE CANAL MEETS ITS WATERLOO

In the landscape surrounding the canal, Wellington Street had long been a bustling crossroads. All types of bridges, from the initial wood structures to the later works in iron, had been built at this location. As a variation to this theme, a tunnel designed under the make-work infrastructures program implemented during the Great Depression was also built on the site. However, problems with contractors delayed the opening of the tunnel until 1935. The idea of creating tunnels under the canal was nothing new. Beginning early in the 20th century, a number of tunnel projects were advanced but failed to receive the assent of the waterway authorities. This refusal was based on the following line of reasoning: digging beneath the canal would hamper any subsequent widening and, above all, any future deepening of the canal. However, in 1921, a board of inquiry on the St. Lawrence shipping route came to the conclusion that industrial settlement had become too dense along the edges of the canal and that it would cost too much to expropriate factories in order to expand it. That is why it was proposed that the route of a new canal be laid out along the south shore of Montreal Island. Construction of this tunnel ultimately spelled the demise of the Lachine Canal and its illustrious career.

Canadian National Railway Lift Bridge over Lachine Canal, National Archives of Canada, Dominion Bridge collection, PA 202868, ca 1940

OVERHEAD CLEARANCE*—TEN STOREYS HIGH

During the second half of the 19th century, three railway bridges were erected over the canal. The Grand Trunk Railway chose Point St. Charles for the location of its building and maintenance centre as well as its switching yard. The presence of railway infrastructure on such a large scale had numerous repercussions on the canal area, in terms not only of increased rail traffic but also of the industrial activity generated by the neighbouring railway shops. The first railway bridge built along the canal was the St. Henri Bridge, completed in 1855. It provided the Grand Trunk with a direct link to Toronto from the Point St. Charles switching yard. Furthermore, this same yard was located at midpoint along the future Toronto/Portland (Maine) line, which was operated by the Grand Trunk following the opening of the Victoria Bridge in 1859 for the purpose of creating a wintertime outlet to the open seas. A second railway bridge, located at the Wellington Street intersection, was erected approximately 15 years later for the purpose of providing access to plants along the north side of the canal. This photo depicts the bridge built on the same location at the onset of World War II. With an apron rising 30 metres high, the bridge was linked to the main railway station only a few years after its completion. The last of the three railway bridges, the Rockfield Bridge located near Lachine and owned by the Canadian Pacific Company, was built in 1885. However, although rail traffic grew steadily throughout this era, a boat always took precedence over a train.

* *overhead clearance*: the distance between the water level and the underside of a bridge.

Around the Locks

Archives Nationales du Québec (Québec), E21, Quebec Department of Lands and Forests, Compagnie aérienne franco-canadienne series, N49-16, ca 1930

AROUND THE ST. GABRIEL LOCK

The St. Gabriel lock sector has one feature distinguishing it from the other two water power sectors. Here a change in ground level occurs on the order of 2.6 metres. As the resulting drop of water accelerates current speed, this declivity lent itself well to laying out hydraulic infrastructures. The current could be diverted from the main canal via mains or side channels, while water could return to the canal on the other side of the factories via a tailrace. This photo shows, along the north side, a portion of the 1825 canal, which was converted into a headrace for supplying water power. It also depicts a tailrace that returned water that had been shunted away from the canal. Only the tailrace, which runs at an angle to the canal, can still be seen today. The same type of installation was also set up along the south side of the canal; there, the tailrace emptied into a basin appearing to the left of the reservoir. Only a portion of the basin and south side tailrace can be seen today. St. Gabriel is the only sector of the Lachine Canal where the banks on both sides were used as sites for hydraulic facilities.

Lachine Canal, St. Gabriel Waste Weir, Looking West, National Archives of Canada, Department of Public Works collection, PA 109917, 1903

THE ST. GABRIEL WEIR

Some industrialists were granted the privilege of diverting water upstream from a lock and then discharging this water downstream from the same lock. Such diversions meant that the upper reach lost a considerable quantity of water to the benefit of the lower reach, which had the dual effect of speeding up the current and slowing down ships headed upstream. The St. Gabriel weir provides a good illustration of the situation. The weir and the lock were located side by side, and the water pouring out of the weir, combined with that of the tailrace, emptied into the canal at a point just behind the location from which this view was shot (in other words, for the ship moving upstream, at the approach to the lock). As a result of the new regulation weirs at Lachine and the change in ground levels occurring from one reach to another, the speed of current rose to more than 4 kilometres an hour, whereas the average speed of a ship in the late 19th century was 5.5 kilometres an hour. Against this backdrop, it is easier to appreciate the complaints of pilots and shipowners, particularly when they encountered such a strong current close to locks where manoeuvres were particularly difficult. Not surprisingly, accidents were a frequent occurrence.

National Air Photo Library, Natural Resources Canada, A-11717-246, 1948

THE CÔTE ST. PAUL LOCK

The Côte St. Paul sector is similar to that of St. Gabriel: the lock site corresponds to a change in ground level of 2.6 metres, which facilitated the development of hydraulic infrastructures. However, in contrast with the St. Gabriel sector, water power facilities were located only along the south side of the canal. This aerial photo shows the 1874-1883 lock on the north side (the longer of the two structures visible), the 1843-1848 lock (the shorter structure), as well as the headrace, which initially runs at an angle to the canal and then parallel to it. Located between the headrace and the 1843 lock, a weir supplied water to a small hydroelectric power station built in 1904, whose output was directed to lighting the canal and operating the locks. The use of water power in the sector continued on into the 1960s, at which time this industrial complex was expropriated by the city of Montreal. For many years, the Côte St. Paul complex was considered distant from the central business district, which delayed its take-off. However, once launched, it stood out for the stability of its factories.

The Ville LaSalle – Ville Saint-Pierre Sector

Canadian Car and Foundry Co., Ltd. (Dominion Plant),
Centre d'archives Hydro-Québec, Shawinigan Water and Power collection (F1), F1 / 700862, 1936

THE VILLE LASALLE – VILLE SAINT-PIERRE SECTOR

This particular photo speaks volumes. At the beginning of the 20th century, the sector encompassing the cities of LaSalle and Saint-Pierre had attracted only a few large plants such as Canadian Car and Foundry Co., Ltd, which can be seen here. The land was for all intents and purposes unoccupied, a fact which encouraged industries to locate here. Obviously, the abundance of available lots contributed to industrializing the sector but it was far from being the only factor involved. To begin with, some of these factories produced dangerous substances, such as manufactured gas or chemical fertilizers, and had to be located outside of inhabited areas. For others, siting had nothing to do with the product they turned out or a lack of space, but stemmed instead primarily from the availability of water that could be pumped directly out of the reach. This question of water was of greater concern to company executives than some contemporary observers are willing to admit, and it became a regular source of predicaments for canal engineers. For example, prior to changing its industrial processes, a factory such as Canadian Tube and Iron, located a slight distance downstream, annually pumped more than 13.5 billion litres of water from the main canal reach extending 7 kilometres between Côte St. Paul and Lachine. Given these needs, it would have been scarcely feasible for this or a similar company to settle in St. Gabriel or Côte St. Paul, considering that engineers already had a hard enough time of maintaining sufficient water in these reaches to meet the needs of industry and the shipping companies.

Jackknife Bridge, Lachine Canal, National Archives of Canada, Andrew Merrilees collection, PA 202368, undated

FROM HOWE TO STRAUSS

The names of Howe and Strauss in question here have nothing to do with the celebrated hockey players or musicians. Except among engineers and a few rare initiates, this title brings nothing very tangible to mind. For the names are in fact linked to bridge architecture. The *decks** featured in the first iron bridges were of the Howe type—that is, they were supported by vertical beams that were themselves maintained under pressure by crossbeams to form rectangularly structured bridges. However, in the case of today's Gauron Bridge, located in Ville Saint-Pierre, the design was altogether different. This drawbridge, or "jackknife bridge," was built according to a design created by the Strauss Company of Chicago and licensed to Dominion Bridge of Lachine. According to newspapers of the time, this bridge was, upon its completion in 1912, the most impressive construction of its type in operation. Featuring a 50-metre-long deck, the bridge could be lifted upwards into a perpendicular position in three minutes, thanks to a counterweight of 725,000 kilograms activated by two electric engines. Today, the Gauron Bridge is flanked by a second drawbridge built in 1959—once again by Dominion Bridge—according to a different design.

* *deck*: the platform of a bridge.

McCarthy Aero Service Ltd., Lachine P.Q., National Archives of Canada, Patent and Copyright Office collection, PA 030760, ca 1920

THE ROCKFIELD SECTOR

As suggested by its name, this sector of the canal was dug directly out of bedrock. It took until the beginning of the 20th century before the canal could be widened to its current width and thereby produce the regular flow required to meet the water needs of industrialists and shippers. This photo provides an idea of the extent to which human and industrial settlement had proceeded over this territory. The appearance of the canal in the local landscape had both a positive and negative impact on Lachine. On the one hand, the opening of the canal caused Lachine to lose its title as the main inland harbour after its function as a point of transshipment had been shifted to a location inside the Montreal city limits. On the other hand, thanks to its industrial siting policy, Lachine was able to attract Dominion Bridge, Dominion Wire, Allis-Chalmers, and Jenkins Valves, among others, to set up facilities on its territory, and the canal proved to be one of the major incentives. For example, until only recently, Dominion Bridge shipped via tow barges that were left moored near the lock products too bulky to be transported by road or railway systems. In addition, all these factories, without exception, pumped water from the canal to meet their industrial requirements. Concerning the land in the Rockfield sector proper, located east of Lachine, many years went by before it began to develop fully.

Canadian Pacific Railway Bridge, Parks Canada, Quebec Canals, 1904

THE ROCKFIELD WALLS

When the canal was widened for the second time (1874-1883), the walls of the longest reach between Lachine and Côte St. Paul were laid out as embankments, at least along the sections requiring such work. Stone extracted during the canal excavation process was used to build up the walls of the section. These masonry stones were small, thus making wall surfaces sensitive to cracking under the impact of the freezing-thawing cycle and prone to crumbling due to the action of the current. After having granted the first contract for rebuilding the wall in stone at the very end of the 19th century, the engineers of the Department of Railways and Canals changed their mind and decided to use concrete in subsequent repairs. In their view, it would take nine years to complete the project of rebuilding in stone, whereas concrete could be poured much more easily and quickly. Beginning in 1905, the walls of the section located between the Canadian Pacific bridge and the Lachine lock were done over in concrete, and the former oak mooring bitts were replaced by cast-iron bollards. Along the walls of this entire reach, this work continued until 1912, requiring eight years in all to complete. During the 1930s, the walls of the other reaches were capped in concrete. As a trace of their presence, some workers occasionally marked the date in the hardening cement.

Lachine

McCarthy Aero Service Ltd, Lachine, P.Q., National Archives of Canada, Patent and Copyright Office collection, PA 030762, ca 1920

LACHINE'S CANALS

Lachine is the only place where all three canals can be embraced in a single glance. In this photograph, a boat can be seen entering the lock built during the 1874-1883 phase. Visible immediately to the north of this lock is the weir dating to 1900, and, parallel to this, the lock constructed during the 1843-1848 period. It is this lock which was at one time filled in and has been recently excavated to be made open to small craft. Beginning in 1926, this lock was no longer used by ship traffic, but it continued to occasionally serve as an additional intake whenever the water level of Lake St. Louis descended too low. The first canal can be seen forming a long ribbon barely 15 metres wide along the edge of the street. A second weir is located on the site of its former lock. A jetty distinguishes the second canal. The two weirs played an important role, for they served to maintain sufficient flow with which to meet the needs of industries and shipping companies alike. For engineers confronted with managing water issues in connection with the Lachine Canal, striking a balance between these competing requirements was an ongoing challenge.

C. A. Barbier Photographic Studio,
Entrance to the Old Canal, Lachine, Musée de Lachine collection, LR-7, undated.

A MAKEOVER

Subsequent to its first widening, the canal was shortened by nearly one kilometre and a half, corresponding to the distance between its mouth and the first lock of Lachine. Henceforth, it measured only 14 kilometres in length, the stretch of the old canal with its bare tapering shores bordering the city of Lachine having been abandoned. This bucolic setting, where children came to play and which was even depicted on post-cards, was nothing more than the first canal,which had lapsed into disrepair in the vicinity of the former Hudson Bay Co. fur trade warehouse. Encroachments and erosion combined to further reduce its width. Then, during the 1930s, the walls of the sector were rebuilt as part of a make-work infrastructure project designed to absorb unemployment created by the Great Depression. At the time of this work, the banks of the canal were delimited by stone and concrete walls, which can still be seen today.

Lachine Canal. Old Canal Showing Wooden Bridge and Old Stone House, Looking East, National Archives of Canada, Department of Transport collection, C 063601, 1907

Water over Wing Dam at Lachine, National Archives of Canada, Department of Public Works collection, PA 110016, 1943

DIVERSIONARY MEASURES

The first wing dam, or jetty, was built during the work to expand the canal that took place between 1843 and 1848. While it has scarcely budged from this position in the years since, the dam has grown larger, particularly after the central portion was significantly widened around 1955. This 1417-metre-long jetty was constructed using stone-filled cribwork rising to one foot above the high water mark. A small lighthouse, the first to be installed upriver from Montreal, was erected at the far end. The jetty built for the third canal was relatively longer and wider than the other one; its pier originally measured 1,890 metres in length. Once completed, the jetty created a new canal 61 metres in width. Wing dams were important for several, if dissimilar reasons. To begin with, shipowners complained about the rapidity of the current in Lake St. Louis. The dam served to arrest the current and, in spring, to break up ice that had formed on this lake. But according to 19th-century engineers, the dam was also used to divert more water into the canal in response to its ever-growing consumption by factories.

Regulating Weir between Locks, Lachine,
Parks Canada, Quebec Canals, 1903

SUPPLYING THE CANAL WITH WATER—THE LACHINE ENTRANCE

Beginning in 1846, the Lachine Canal took on a threefold function as the industrial water and energy supply system was combined with the navigable waterway system. For years to come, this triple mission was to be a continual source of headaches for engineers. The 1825 lock already played a regulating role, allowing sufficient quantities of water to enter and supply all the reaches of the canal and offset certain losses of water caused primarily by leaks, evaporation, and lockage operations. The arrival of factories with an unslakable thirst for water only served to complicate the calculations of engineers. From then on, they would have to factor in not only the shifts in water mass represented by locking operations, but also to assess the water consumption of the new plants in terms of power and process needs—and, above all, to evaluate the shifts in volume entailed by such increased use. At issue, fundamentally, was water management, and it was at Lachine that these management measures were begun. Lachine was where water entered the canal and became, several years later, the site of both regulating weirs, with the first one resulting from the conversion of the former 1825 lock and the second being built beginning in 1854. The weir built using the former lock was expanded in 1870, with 16 sluice gates being added to its structure. In 1900, the weir dating to 1854 doubled in size, with the number of discharge structures also doubling. It is this weir which is shown in the photograph.

Bridge at Lachine above New Lock No. 5, Lachine Canal, Looking West, National Archives of Canada, Department of Transport collection, C 081815, 1903

VARIATIONS ON A THEME

Prior to the start of construction of the canal, legislation prescribed that if a canal eventually cut through a public road, a bridge would have to be built in less than one month and that any existing bridge would have to be replaced. Thus, bridges became an integral part of the local landscape from the earliest days of the Lachine Canal. In 1841, it was spanned by 14 bridges, including two built in cut stone, with one being located in Lachine and the other near the next-to-last lock in Old Montreal. All the other bridges were wood constructions. However, when the canal was widened between 1843 and 1848, these older stationary structures had to be done away with. As steamships had now begun to travel up and down the canal, these infrastructures were replaced by wood *pivot bridges*.* Whereas Durham barges could be unmasted and were able to accommodate towing, the steamships were outsize for the overhead clearance of stationary bridges. As a result of such changes, a new category of employees, the bridge operators, soon cropped up in the inventory of local jobs. The pivot bridge shown above was completed in 1882-1883; its machinery was activated by electricity around 1909. This bridge was replaced with a steel construction in 1922 by Dominion Bridge, a nearby company with a continent-wide reputation. All in all, this sector of the canal would be home to four different types of stationary or moveable bridges.

* *pivot bridge*: a bridge whose deck can be rotated about an axis consisting in a pier built in the centre of the canal bed.

Workers of the Canal

Lock House North of St. Gabriel on the South Bank of the Weir, National Archives of Canada, Department of Public Works collection, PA 109970, 1923

A ROOF OVER THE HEADS OF LOCK MASTERS AND BRIDGE OPERATORS

The workers who looked after locks 16 hours a day had to find shelter from inclement weather—for there was a lack of lodgings in the vicinity of the locks—especially when boats were a long time coming. From the time of the first canal, huts for lock masters began dotting the landscape around the locks. In the 25 years after they first appeared, these uninsulated wood shelters used as living quarters for the lock masters' families had become completely outmoded and unsanitary. That is why, during the first canal widening phase, engineer Barrett recommended rebuilding these makeshift structures so that lock masters and their families would be better protected from wind and rain. Beginning in 1855, part of a lock keeper's wages was paid in the form of an allocation of firewood or even a patch of pastureland on which to graze his cows, but such measures disappeared during the last decade of the 19th century, when lock masters and bridge operators began receiving an annual allowance with which to pay their rent. Since the first bridges were stationary, no one was needed to handle the duties of a bridge operator. But as pivot bridges began to emerge and bridge operators were hired, shelters for these employees necessarily followed, becoming a part of the local landscape until the canal was finally shut down.

Works on the Lachine Canal, Bibliothèque centrale de la ville de Montréal, ca 1875

OCCUPATION: ENGINEER

They went by the names of Thomas Burnett, Alfred Barrett and John Page, but all three were alike in that they were engineers, or at least served as such, and had signed the plans of the three different Lachine canals. Burnett, born in Scotland, arrived in Montreal on the recommendation of Thomas Telford, the celebrated British canal engineer. Burnett designed the plans of a narrow canal in the English style, producing a canal that was closer to the kind used for barges than one intended for boats. Barrett, who was born in New England, made his engineering début on the Erie Canal. His career later took him back and forth across the border overseeing various canal projects, including work on the Welland Canal. On the basis of his report, in which he painted a rosy picture of the considerable gain to be had from leasing hydraulic lots along the Lachine Canal, the Department of Public Works launched into the adventure of hydraulic power in 1846. As Barrett's calculations proved overoptimistic, Page was the one who had to attempt to correct a situation that had become distinctly difficult, to say the least. Page, the head engineer of the Canadian Department of Public Works, only signed plans and endorsed the decisions of his subordinates. Most likely, these decisions were those of John Sippell, the resident engineer at the Lachine Canal. Whatever the case may be, both men had to remedy a predicament that was Barrett's doing and was itself merely a reflection of the first two engineers' inadequate command of hydraulics principles.

Côte St. Paul, North Side, National Archives of Canada, Department of Public Works collection, PA 119069, 1909

A PERPETUAL WORKSITE—AND A TEST BED FOR FUTURE BUILDING CONTRACTORS

Who were the contractors responsible for building the canal? Of the four charged with completing the first canal, only White and Phillips had experience in construction, while the other two, Bagg and Wait, were merchants and businessmen. During the first widening project, Mason and McGrath, the contractors who oversaw most of the work, were also well-known Montreal merchants. This was consistent with the North American approach to canal building, whereby contracts were awarded to merchants or professionals who sought profits proportional to the scale of the planned works. But this approach was not always successful, since numerous contracts were cancelled mid-project. Thus, it took until the second phase of canal widening before Montreal saw the emergence of a class of building contractors who travelled from worksite to worksite with their own employees. Such was the case of James Worthington, a native of southern Ontario who, prior to receiving the contract to build the first two sections of the canal, had completed several rail projects in southern Ontario, raised two of the three forts in Lévis comprised within the city of Québec defence network, and finished work on the Intercolonial railway in the Maritimes. From that time on, major building or repair contracts were awarded to prominent contractors such as Quinlan and Robertson.

Old Basin No. 1, Parks Canada, Quebec Canals, 1903

WALK OUT—BUT WHERE TO?

The worksites of all three canals were paralyzed by ten different work stoppages. In these conflicts, it is possible to glimpse the evolving dynamics of workers' struggles. Initially consisting in a mere refusal to work, conflict degenerated into riots in 1843. In 1877-1878, certain groups of workers assembled in unions also walked out. In 1825, carpenters, many of whom had been recruited from among recently arrived Irish immigrants, demanded better wages, thus following the lead of the masons who had struck two years earlier. The reply forthcoming was categorical: either accept their lot or find work elsewhere. During the 1840s, canal building in the United States went into a major slowdown, with the result that several thousand workers migrated from one worksite to another and eventually came to Montreal seeking jobs. Profiting from this abundant supply of manpower, a subcontractor by the name of Mason cut salaries, upped the number of working hours, and delivered pay (which, obviously, reflected only the hours worked) in the form of coupons redeemable in his store. As a result, few workers had enough money to survive on. Numerous workers, disgruntled with the store's usurious practices, rose up in rebellion. The army intervened on behalf of the contractor—which comes as little surprise, considering the interests of the government in these construction projects. Thus, the stage was set for the appearance, during the 1877-1878 work phase, of unions representing skilled workers such as masons and stonecutters. While members of the skilled trades were able to make their claims prevail twice, such was not the case with day labourers, whose own attempts at gaining better conditions failed four times. At best, they received partial satisfaction, at worst they had to find work elsewhere.

Interior of a Dredge, Parks Canada, Quebec Canals, 1908

STATUS: LACHINE CANAL EMPLOYEE

An 1893 register of canal employees lists 113 individuals, 95 of whom were assigned to operating bridges and locks. Nine out of every ten of these workers were seasonal employees hired for periods varying from five to seven months. Bridge and lock workers received the lowest monthly wages, after telephone operators. Topping the pay scale was, obviously, the resident engineer and the electrician, who owed his prominence to the novelty of his duties. Next on the list were the wharfinger and his assistant, who were responsible for monitoring and coordinating unloading times. Further down in the register came the boom master and his assistant, who positioned booms in order to prevent stray logs and other debris from clogging water intakes. Thus, in the late 19th century, by far the most sizable groups of canal employees were those actually involved in canal operations. The situation changed somewhat during the following century; during the 1920s, the canal workforce doubled. From then on, canal employees belonged to one of three groups: office, operations, or maintenance. Hiring would be maintained at this level until the closing of the canal in 1959. By that time, lock masters and bridge operators accounted for no more than 60% of the employees, as a result of the electrification of operations implemented at the turn of the century. Maintenance workers such as forge operators, carpenters, and electricians accounted for close to one third of the staff. Serving as a com plement to this workforce was a crew of dredgers, a dozen-odd individuals who were assigned to canals throughout all of Quebec.

Lachine Canal, National Archives of Canada, Department of Public Works collection, PA 110241, undated

DREDGE OPERATORS AND DIVERS

Beginning in 1849, the waterway authorities engaged crews of dredge operators and divers, whose duties often complemented one another's. In the realm of canals, dredge operators are indispensable, for they are responsible for eliminating shoals on which ships might run aground. In the case of the Lachine Canal, whose current was particularly rapid, it did not take long for silt to accumulate. And, since engineers could not always assure shippers of maximum depth on account of the overconsumption of water by factories, shoals constituted a serious source of problems. Once loaded onto barges, the silt was carried to a particular sector, the former Lac à la Loutre, and discharged. This sector, which lay below the level of the canal, had only one clay backing by which canal walls were kept watertight. Depositing fill at this location thus served an important purpose. During the 19th century, divers were needed to closely search the floor of the canal for stones and pieces of sunken wood, or to check on the state of canal walls. Since a rock or bit of debris could interfere with the opening or closing of a gate, these men saw duty until navigation on the canal ended in the 1960s.

Floods

Flood of St. Rémi Street,
St. Henri Historical Society,
negative 42-ph-1, 1915

SPRING FRESHETS

Spring, with its sudden thaws and its ice jams, was greeted with a sense of foreboding by canal operators and the neighbouring factories, especially in the Montreal sector. Obviously, such devastating floods as those of 1861 and 1869 (followed by low water levels in the summer), or the 1886 flood, which was remembered as the most extensive inundation of the 19th century, were perhaps understandable in the time prior to the construction of the Mackay jetty and the flood wall on de la Commune St., which were to become well-known Montreal fixtures. However, such absences do not explain why, in 1915, the residents of St. Rémi Street in the St. Henri sector found themselves flooded by several centimetres of water. Thus, the notion that flooding only occurred in the harbour area must be viewed with some suspicion. The canal could also represent a "wellspring" of floods. As early as 1824, several canal-side landowners complained that their land had been flooded on account of a leak in the system. However, canal engineers suffered more headaches in connection with the St. Pierre River culvert, which regularly overflowed and inundated the neighbouring lands, to the great distress of the operators of the Grand Trunk, whose railway line bordered the canal in this area. In 1896, the city of St. Henri lodged a formal protest with the federal government, enjoining it to settle this problem once and for all. According to the affidavit served, the St. Pierre River received sewage from the city of Lachine in addition to the discharge of various industries located upstream, and the culvert was unequal to the task of handling this situation. As a result, the houses lining the Côte St. Paul Road were regularly flooded. An effective solution to this problem would only be found in the 1920s.

Shipping and Fleets

CARON, Paul, *Flour Mills on Lachine Canal*, National Archives of Canada, C 148390, ca 1920

A network of canals like the one developed along the St. Lawrence can be compared to a modern highway system. Of course, there is no forgetting that the St. Lawrence/Great Lakes waterway is the longest inland navigable route in the world, covering a distance of 3,800 kilometres, of which canals account for 114 kilometres. It carries as many passengers as it does goods, and the fleets belonging to shipowners and shipping companies are very much like fleets of trucks and buses. Furthermore, traffic on both systems is regulated by rules of the road. Shipping companies and shipowners are driven by the same economic imperatives—turning a profit on travel in both directions and reacting to competition, whether it comes from the railway or the Americans—and to do so, they must regularly replace and upgrade their fleet. Only one major difference separates the waterway from the highway—namely, the need for transshipment, which caused silos, marine towers and, inevitably, wharves to become part of the landscape.

Bulk shipment

Competition failed to materialize between American merchants and their Montreal counterparts, who were keen to appropriate the grain trade from the interior of the continent and thereby compete with the Erie Canal. By the 1850s, the disproportion between traffic on the Erie Canal and the St. Lawrence was enormous, as less than 10% of wheat exports were carried by the St. Lawrence system. In the late 19th century, the situation remained unchanged, since barely 13% of wheat transited through the St. Lawrence. A number of factors accounted for this situation.

In the mid-19th century, the wintertime closing of the St. Lawrence River clearly handicapped Canadian maritime transport, whereas in the United States, canal barges and trains provided year-round access to the port of New York. Certain shipowners also pointed either to the lack of transatlantic transportation and the high rates charged for freight, or to competition from the Grand Trunk Railway. While it is true that in Canada trains gained the favour of flour millers for their shipping requirements, ships were nevertheless better suited for bulk transportation. In the late 19th century, winter was still as much a problem as ever. Despite the hundreds of millions of dollars invested by the Canadian government to modernize the country's canal system, at heart the problem remained the same as before: the waterway formed a bottleneck, since ships sailing on the Great Lakes far exceeded the dimensions of canals. As a result, grain—American grain more so than Canadian—was shipped to Kingston, where it was transferred onto barges. From there, the barges were towed to Montreal, thus increasing transportation expenses, as grain again had to be transferred, this time onto ocean-going vessels.

In 1903, the canals were standardized to eliminate the transshipment stopover in Kingston. The development of grain growing on the Canadian Prairies, the quest for markets for wheat, and the return to prosperity after the international recession ended in 1896 all combined to create the impetus necessary to expanding Canada's merchant marine. From 1902, ships began carrying greater quantities of Canadian wheat. Whereas the fleet had consisted of 27 grain ships in 1899, shipowners acquired more than 120 such vessels over the next 15 years, most of them steel-hulled freighters having dimensions in line with canal clearance requirements and yet capable of navigating on the Great Lakes.

Thus, boats changed dramatically in the early 20th century. Until that time, the Great Lakes ships had been schooners for the most part. This type of vessel had been adapted to lock dimensions, thus earning them the nickname of "canallers." When, some time later, these sailing ships were stripped of their mainmasts, they completed their life cycle as barges. They also became more profitable, since by this stage in their active life, they usually were entirely paid for. As a result of the new lock widths, and considering the particularities of the Cornwall lock, the maximum length of the canaller was 77.27 metres, and maximum width 13.28 metres. Nevertheless, from a previous deadweight capacity of 400 tons, ships could now transport six times more cargo. The 20th century saw the canallers become diversified and specialized.

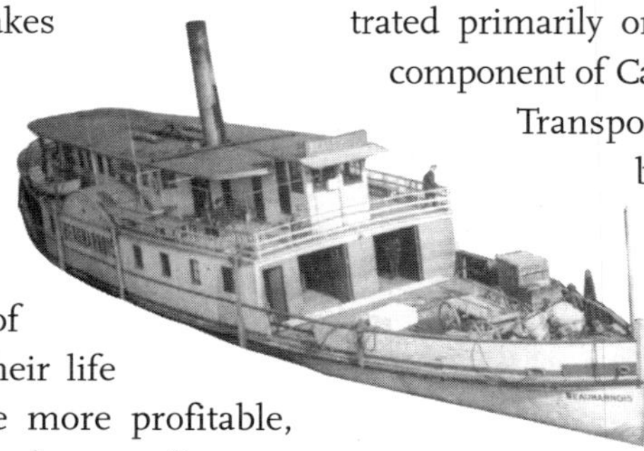

Shipping Lines

Whereas during the 19th century, shipping companies were modest affairs most often owned by small groups of investors, the early 20th century witnessed the birth of one of the largest Canadian inland shipping companies—Canada Steamship Lines, founded in 1913. The result of the merger of several shipping companies into a single outfit, it was part of a pan-Canadian and even a North American and British trend, particulary in the manufacturing sector. Canadian examples of this movement include Dominion Textile (1905), Imperial Tobacco (1908), Canadian Car and Foundry (1909), Canada Cement (1909), Stelco (1910), and Matthews-Laing (1913). All of these companies had one or even several mills along the Lachine Canal.

Canada Steamship Lines brought together 10 shipping companies, some of which were the product of mergers themselves. Some of these firms, such as the Richelieu and Ontario Navigation Company, concentrated primarily on passenger transportation. A later component of Canada Steamship Lines, the Montreal Transportation Company, had interests in bulk transportation and wheat shipments. Both companies had their headquarters in Montreal. But with the growing interest in the Western wheat trade, several companies settled in Ontario on the shores of the Great Lakes. In addition, the Canada Steamship network extended as far as Great Britain, since representatives of Furness Withy and Vickers also had a hand in guiding its fortunes. By pooling the activities of these various shipping companies, more than a hundred ships were brought together under the banner of Canada Steamship Lines.

All told, more than 50 shipping companies were registered along the Lachine Canal between 1825 and 1945 as leaseholders of either wharves or warehouse space. Although these companies were Canadian for the most part, a number of American companies such as the *Chicago Tribune* also had their own wharf for several years. However, this did not mean that Americans were the only "outsiders" who travelled up and down the network of the St. Lawrence canals. For by the late 19th century, European ships were also using the

Lachine Canal to travel as far inland as the Great Lakes.

The vessels of Canada Steamship Lines and its subsidiaries did not limit themselves to carrying wheat, because that would have meant "travelling light" (that is, with an empty load) upbound to the Great Lakes. Although west-to-east trade predominated, a ship could nevertheless carry grain to Montreal, continue on its way to Nova Scotia, and then return with a cargo of rails. However, as of 1906, iron ore began to figure among the cargoes shipped to the Great Lakes area, as did pulpwood, although in this case, an embargo placed a limit on the quantities that could be shipped to the United States, where Chicago was the main destination.

The Harbour Landscape

Such hustle and bustle was bound to produce an impact on life around the Lachine Canal in terms of traffic and the landscape. In particular, the standardization of locks throughout the entire St. Lawrence system had the effect of stimulating the wheat trade. For several decades previous to this improvement, Montreal possessed certain infrastructures for receiving and storing grain, such as the Montreal Warehousing facilities and elevators along canal reach no. 2 and in the harbour area. But as upgrading was completed, the quantities of wheat shipped to Montreal rose dramatically, to the point that port authorities decided to set up a new grain silo on the edge of the Lachine Canal. It was called Elevator B, now known as Silo no. 5, a monumental steel structure that was quickly neighboured by a second, concrete elevator.

The various basins built in the middle or the last quarter of the 19th century became full-fledged wharves and were covered with hangars or warehouses leased in whole or in part to various shipping companies. These basins, which were built along canal reach no. 2, nevertheless preserved the uses that had been assigned them at the time of their implementation. To the east of the Wellington Bridge were basins 6 metres deep, capable of receiving ocean-going ships, and immediately to the west of the same bridge were basins 4.5 metres deep, reserved for inland navigation.

Wherever there were wharves and warehouses, there were also sure to be longshoremen. Some of these men were employed by docking companies while others were associated with various canal-based companies that had their own wharves and unloading infrastructures. During the 19th century, dockers were driven to strike on account of their working conditions, but during the 20th century, this tendency appears to have subsided, perhaps due to general economic conditions, whether favourable (World Wars I and II) or unfavourable (the Depression). The situation certainly did not stem from the marked drop in transiting ships, as paradoxical as this may seem.

Whereas an average of 10,500 ships passed through the canal annually between 1870 and 1900, this figure dropped substantially, by close to 20%, as part of the downward trend occurring throughout the 20th century. However, freight tonnage carried continually rose during this period, thus indicating considerable growth in ship size. Not all these ships were bulk carriers or steamships—not until the early 20th century at any rate. Since the mid-19th century, tugboats had been towing sailing ships or barges, while freighters travelled alongside passenger ships and packets, a kind of rapid cargo ship that carried goods of smaller dimensions and mail. Although plying the same waters as freighters, the passenger ships offered either general transportation services or tourist cruises.

This proliferation of ships did not fail to generate a certain amount of confusion and congestion on canal waters, thus compelling the enactment of a series of navigation regulations which were periodically revised in keeping not only with the evolution of river fleets but also with the advent of new technologies. The use of telephone and electricity, in particular, made it possible to check whether cruising speed was complied with and also to accelerate lockage operations or pivot bridges in order to give ships the right of way.

Furthermore, such numbers of ships inevitably increased the accidents that occurred on the canal—be this a bashed-in lock gate, a collision between two ships, or the grounding of a boat in the waterway. Regardless of whether the accident involved an ocean-going vessel or an excursion boat, all repairs could be performed on location since the canal had no fewer than five dry docks. Furthermore, these same dry docks were associated with shipyards that built ships which subsequently travelled up and down the canal and the river, at least during the 19th century.

Despite all of this teeming activity and the millions of dollars swallowed up in modernizing the canals, the original problem remained intact, with the result that cargo carried on the St. Lawrence system attained a plateau. Total bulk transiting on the St. Lawrence canals peaked at 11,000,000 tons at the very end of their active life. Furthermore, whereas ships travelling along the St. Lawrence canals could carry loads of up to 2,500 tons, vessels on the Great Lakes carried ten times more. In a way, the canaller represented one link in a longer chain. However, the time had come to take corrective action, which took shape in the St. Lawrence Seaway, inaugurated in 1959. From then on, it was out of the question to lose two hours travelling up the Lachine Canal. Commercial imperatives again hove into view.

Transported from the 19th into the 20th Century

Canal Fleets

Timber Raft in Lachine Rapids, McCord Museum,
Notman Photographic Archives,
MP 208 / 75, ca 1900

TIMBER RAFTS ON THE CANAL

Although timber rafts were allowed into the canal from the time it was opened, raftsmen preferred to shoot the rapids because that way they could avoid paying the toll. Most of the time, these timber rafts consisted of squared lumber sent down river to the coves of Québec. Care had to be taken assembling the rafts of fuel wood cut on the newly settled lands in the Ottawa River valley; obviously, rafts were not allowed to create a jam in the canal. Any stray or half-submerged log had to be removed. However, scarcely three years after the opening of the canal, the commissioners recommended banning these rafts because of the damage to shore embankments. As of 1831, a ban was declared on the floating of rafts down the canal any time after August 1 of each year. Beginning in 1848, the authorities attempted once again to regulate lumber convoys by making it unlawful to unload them on the same side of the canal as the towing path and by requiring this operation to be carried out within 24 hours of the convoy's arrival. However, even as recently as 1907, timber rafts were still permitted on the Lachine Canal as long as they had been inspected beforehand. Raft dimensions were not allowed to exceed a maximum of 8.5 metres by 97.5 metres, and had to be divided into two sections of 48.75 metres each. In later times, wood was carried by barge even as it was being gradually supplanted by coal as a fuel source.

Redpath Sugar Archives, 1897

SAILING SHIPS AT THE CANADA SUGAR (REDPATH) DOCK

By the end of the 19th century, "batteaux" and Durham barges had long since ceased to ply canal waters. Used primarily on the first Lachine Canal (constructed between 1821 and 1825), such vessels were better suited to the difficult sailing conditions prevailing in the early 19th century. These flat-bottomed boats displaced up to 40 tons and had a *draught** of no more than 75 centimetres. However, with the subsequent widening of the canals and, above all, its deepening first to 2.75 metres and then to 4.25 metres, sailing ships were adapted to the St. Lawrence system. They evolved into the canallers that were built specially to meet the limitations of locks. These wooden two- or three-masted schooners were rigged out with square sails that enabled them to take fuller advantage of the prevailing winds when travelling downstream. A typical sail canaller measured 53 metres long and displaced 350 tons. This class of ship sailed up and down the St. Lawrence and its canal system until the end of the 19th century; the last one disappeared only in 1905. The ship seen moored at the far end of the Redpath dock is one of these sail canallers, while the other is an ocean-going vessel that must have been shallow, as it had succeeded in making its way through the ocean basins.

* *draught*: the distance extending vertically from the float line to the keel.

Wm. Notman & Son, S.S. *Turret Crown in Lachine Canal Locks*, McCord Museum, Notman Photographic Archives, II-111874, 1895

STEAMERS OF ALL SORTS

How did steamers differ from one another? Let us count the ways. First of all, there was a distinction between freighters and passenger carriers, even though both classes of ships were built as side paddle wheelers or stern wheeler steamers. But other types of boats travelled up and down the canal as well. The first screw-driven ships appeared in the 1840s. This technology had considerable advantages, since in particular it enabled ships to take up the entire width of a lock. When these boats were adapted to the dimensions of the locks in the third navigation network, they were nicknamed "canallers." However, such vessels were not designed to be fast because any potential gains in time would be quickly lost in lockage operations. Canallers did not all have the same type of holds, since internal architecture varied depending on the type of cargo to be carried. Thus, the 20th century saw the emergence of coal ships, cement ships, oil ships, and pulp and paper carriers. Another type of vessel—the turret ship— also enjoyed a short-lived existence on the St. Lawrence canals. Of British make and having a peculiar, submarine-like shape, the turret ship was introduced in Canada by Dominion Coal of Nova Scotia for carrying coal as far inland as Montreal. Once their charter contracts had expired, these ships were used on the canals and the Great Lakes. Owing to the particular shape of the vessels, their owners could play off tonnage regulations in order to pay lower tolls. Around 1911, ships driven by diesel engines appeared for the first time, but they only became popular after World War II.

Close-up of the Montreal Harbour Commissioners' 75-Ton Floating Crane Lifting a Car of Coal, Bibliothèque nationale du Québec, Pharon collection, 1385, undated

BARGES, SCOWS, AND PONTOONS

"Barge" can be a confusing term where canal boats are involved. Some might be inclined to equate this word with the "batteau," the Durham barge, or the engineless flat-bottomed boat used in the late 19th century. It should be noted that all of these boats were flat-bottomed, but that each was used during a different period. The batteau, which saw service during the late 18th century, was ideal for the fur trade. It could hold up to four tons of goods and was akin to a dory fitted with a sail. The Durham barge, also rigged with sails, measured up to 17 metres in length and could hold a cargo of up to 40 tons. Architecturally speaking, the engineless barge of the late 19th century was like a former sailing ship that had been demasted, although it could also be rigged out with a small sail. Most often, the engineless barge was towed by steam tugboats. The holds of these barges could contain up to 650 tons. Such vessels were greatly favoured along the canals of the St. Lawrence prior to 1904, given that the waterway system had not been standardized until then and products like wheat had to be transshipped in Kingston or Prescott, Ontario, if this canal system was to be used. Standardization of the system, once completed, spelled a quick end to barges in the canal landscape, since bulk carriers could now reach Montreal directly. Finally, the pontoon was a flat hold-less vessel that appeared in the canal during the 19th century and that could be described as a floating platform. As this type of craft had no hold, it was not intended to take on cargo but was instead used to move objects or masses. It should not be confused with a scow, which had a rounded bow and could be used to carry cargo over short distances.

Topley, William James, *At the Mouth of the Lachine Canal*, National Archives of Canada, William James Topley collection, PA 026267 ca 1890

TOWING

Beginning with the first widening phase of 1843-1848, the question arose of how to push traffic through the corridor. Should priority be given to using steam towboats or should the animal-drawn system of towing continue as before? Since work was still in progress, a moratorium was declared on the use of towboats. In 1849, Jacques-Félix Sincennes, one of the shareholders of the Richelieu Company, formed an alliance with William McNaughton, a sawmill operator in the Ottawa valley, and founded the Sincennes/McNaughton Company for the purpose, notably, of providing towing services in Montreal harbour and the Lachine Canal and also for towing timber rafts down river as far as Québec. Beginning in the 1850s, this towing system became the norm, and, during the same period, the service was expanded to include all shipping on the St. Lawrence River from Lachine to Kingston. As is clear from the example of the Sincennes/McNaughton Company, it was private enterprise that provided towing service. With the proliferation of barges and scows, the system spread to such an extent that, in the early 20th century, specific regulations were enacted. From that time on, no towboat was authorized to travel through the canal towing more barges than could be accommodated during a single locking operation. As for rudderless scows, subsequent regulations specified that the distance between the scow and the towboat had to be less than 30 metres. The Sincennes/McNaughton Company was later absorbed by McAllister Towing, which continues to be in operation.

S.S. Hamildoc *at Lock No. 1, Lachine Canal,*
National Archives of Canada, Andrew
Merrilees collection, PA 202576, 1937

THE GIRTH OF BULK CARRIERS

With the passing of the years, ships grew bigger and bigger. To begin with, it should be noted that both sailing ships and steamers sailed the lower St. Lawrence until World War I. However, in terms of through traffic, the balance between steamers and sailing ships was upset in the early 20th century: as of 1904, steamers travelled more frequently over the canal network than did sailing ships. Among the Canadian fleet (which accounted for two-thirds of all bulk carriers on the upper St. Lawrence), sailing ships were smallish and generally displaced between 50 and 200 tons. As for steamships, it appears that during the 1850s and the 1860s builders modelled them on sailing ships as most steamers had a deadweight capacity of 50 to 200 tons. However, beginning in the 1870s, a distinction grew ever clearer: steamships were either altogether small—that is, less than 50 tons—or big—that is, more than 250 tons. And they continued to grow bigger, as the proportion of steamships of 250 tons rose from 6% in 1857 to 66% in 1919. Thus, the *Hamildoc,* which is seen here sailing past Elevator B (Silo no. 5), was typical of ships from the 1920s. Built in England in 1927 for Paterson Steamship, it was 77 metres long and 13 metres wide (whereas the locks of the period measured 82.25 metres by 13.7 metres), and displaced 1,926 tons.

Sail Boats and Tugs at Coal Docks,
McCord Museum, Notman Photographic
Archives, 106959-II, 1894

ENSIGNS, FOREIGN AND DOMESTIC

For the most part, the ships sailing along the St. Lawrence canal system were Canadian, and few American or foreign ships actually used the Lachine Canal. Around 1867, the proportion of ships sailing under the American flag hovered around one third, the direct consequence of the signing of the Treaty of Reciprocity of 1854, which granted Americans the right to navigate on the St. Lawrence. The U.S. Civil War scarcely influenced this proportion, which held steady until the end of the 19th century. During the first two decades of the 20th century, the American fleet's share of traffic tapered off somewhat, settling at approximately 28%. Until the end of World War I, most of the American

ships were sailing vessels, whereas, beginning in 1887, the Canadian fleet consisted primarily of steamers. However, Canadian and American ships were not alone in plying the river system, as several European ocean-going ships specifically designed for canal networks regularly headed for the Great Lakes, beginning in the late 19th century. After a momentary hiatus brought on by World War I, the European ships reappeared in the canal during the 1930s, for the most part hailing from Norway or Holland. They were, however, few in number—that is, less than 5% of all ships recorded as passing through the St. Lawrence canals.

Shipping Regulations

Boats Battling Ice, Lachine Canal, National Archives of Canada, Andrew Merrilees collection, PA 202881, 1933

THE SHIPPING SEASON

Some years, such as 1831, an early warm spring made it possible to open the canal to shipping by April 10, whereas in 1877, the season began as late as May 7. The earliest closing date on record occurred on November 17, 1844, whereas files show the latest closing date occurring on December 17, 1918. The canal was thus always closed to navigation during the winter. Between 1836 and 1958, an average season lasted 222 days. While this average represents an indicator, it can be deceptive, since it glosses over an important phenomenon—the lengthening of the shipping season, beginning in the spring of 1913. Whereas the season had previously lasted 216 days, it was extended to 232 days that year: consequently, shipping began a week earlier in spring and ended a week later in fall. The shipping season was lengthened due not only to the use of steel hulls but also to the increasing power of ship engines. In short, the canal fleet changed and improved, and, as a direct offshoot of this trend, the tonnage of products transported increased considerably.

Triggs, Stanley, *Lock No. 5, Lachine Canal*, McCord Museum, Notman Photographic Archives, 1968, MP 2361-37

WORKING HOURS ON THE CANAL

While at the end of its active life, the canal was open to shipping seven days a weeks, 24 hours a day, such had not always been the case. At the time of its opening in 1825, normal working hours were six days a week from 4:00 a.m. to 8:00 p.m. between May 1 and September 1, and from sunrise to sunset for the remainder of the season. However, subsequent to the first phase of canal widening (1843-1848) and an increase in shipping traffic, several shipowners lobbied to make the canal open to navigation on Sundays. In 1864, a first breach in regulations came when, "in order to promote trade," ship captains were authorized to continue upstream through the Lachine lock Saturday night until 2:00 a.m. on Sunday morning. In addition, the canal was open every Sunday in November until the end of the shipping season. Seventeen years later (i.e., following the third expansion phase), the ban on Sunday sailing grew yet weaker as ships were now authorized to navigate prior to 6:00 a.m. and after 6:00 p.m., with the November schedule being maintained as before. Beginning in 1882, the canal was open day and night. However, only in 1924 did the canal become accessible to ships at all times.

View of Canal from Rockfield Bascule Bridge, Looking East, Lachine Canal, National Archives of Canada, Department of Public Works Collection, PA 108741, 1914

GO SLOW—CANAL CROSSING AHEAD

The first bona fide shipping regulations appeared in late 1827, two years after the canal was opened to navigation. At that time, the horses used for towing were not allowed to go faster than a walking gait—that is, 2 km/hour. Thirty years later, this speed was actually clocked, which meant that if a ship cleared the canal ahead of schedule, it could be held up at the last lock until the time allotted it had been spent. In 1895, steamers had to respect the pace of slower ships being towed, although in any event the customary speed was not allowed to exceed 6.4 km/hour on reaches or in the canal entrances. Beginning in 1908, towing by animals was banned, a development that practically coincided with the end of the era of sailing ships. As a sign of shipping trends, the speed of ships rose to 9.6 km/hour in 1924.

Tug Katherine K *with Scow* Oka No. 10 *and Steamer* Lady Grey *in Lachine Canal*, National Archives of Canada, Andrew Merrilees collection, PA 202649, undated

LOCKAGE PRIORITY

From the beginning of navigation on the canal, a set of rules establishing who had the right of way was set out. Accordingly, timber rafts had to yield to boats, shifting to the side opposite from the towing path. As shipping traffic grew, steamers gained priority over sailing ships. As of 1857, vessels transiting through the canal were ranked according to lockage priority, with mail ships and passenger ships coming first. Furthermore, ship traffic would always take precedence over rail traffic, at least starting in 1907. As mail packets were gradually retired from service during the 20th century, a subtle distinction became detectible; as of 1924, mail boats began to jump the line, so to speak, by registering as government ships, which since the beginning of the 20th century ranked first for right-of-way. In second place came passenger ships with fixed schedules, excursions boats (a new category), and ships carrying perishable goods for market. Ranking last for right-of-way were freight ships subject to a regular schedule, followed by all other freighters plus pleasure craft (another new category). In short, priority was granted the fastest ships operating according to fixed schedules.

NOTMAN, William, *S.S. Turret Crown*, McCord Museum, Notman Photographic Archives, 111872, 1895

LOCKAGE OPERATIONS

As the number of steamers on the canal rose ever higher, the administrators of the canal system were forced to enact regulations governing lockage operations. The first recorded lockage directive dates to 1854, when the engineer Sippell ordered that all ships approaching a lock had to stop 90 metres short of the gates and wait their turn to dock according to an order of precedence. A half century later, this operation had become considerably more complex in structure. A red lamp lit up to indicate when the lock was ready for the ship to enter. Eight hundred metres in advance, ships' captains had to take the precaution of sounding a horn or ringing a bell. Ships had to steer their way carefully through the gate, which had to be completely opened, and then the sill, at which time the screw propeller had to have come to a full stop. In 1921, as traffic became heavier, captains were advised to give warning 300 metres prior to their arrival, sounding the horn not just once but three times, so as to eliminate any confusion. As the ship entered the lock, it had to advance slowly enough that it could be handled only with mooring lines. It was an operation that became all the more critical during the 20th century, since ship forms so closely hugged lock contours that water was unable to flow easily past the stern. Thus, upon entering the lock, the captain had to go "full ahead" in order to penetrate the lock. This highly sensitive manoeuvre resulted in numerous accidents.

S.S. Simla *in Lock No. 1, Lachine Canal,*
National Archives of Canada, Andrew
Merrilees collection, PA 198564, undated

SHIP EQUIPMENT

In the mid-19th century, any ship using the canal had to be equipped with two good mooring lines and a horn. Since vessels were becoming steadily bigger, any ship of more than 200 tons now had to be fitted out with not two but four mooring lines—one fore, two amidships, and two aft. A screen had to be set over steamer chimneys in order to prevent flying embers from escaping. In 1885, ships were required to display a scale indicating their draught, and, two years later, two such scales had become mandatory, with one each at the bow and the stern. Nighttime navigation on the canal began only in

1882, but 25 years before, shipping regulations specified that every ship had to be lit at both ends. Like many other canallers of its kind, the steamship *Simla*, built in Kingston by the Calvin Navigation Company in 1903, featured a *jackstaff*,* at the tip of which a light was hung. This particular set-up stemmed from architectural considerations as well as the location of the pilot house: thus the staff provided a reference point enabling the ship's captain to steer his ship.

* *jackstaff*: a long upright pole located on the bow.

Canal and Canadian Pacific Railway Bridge,
Parks Canada, Quebec Canals, undated

DOWN THE TOWING PATH

In this photo of the Canadian Pacific railway bridge spanning the canal a short ways downstream from Lachine, one particularly noteworthy detail is the towing path. Boats continued to be towed until the early 20th century. The towing path, which measured about 3.6 metres wide, ran parallel to the canal. A ditch bordered it for draining off rainwater. The path was used by a team of four horses that hauled boats from one reach to the next. This service was provided by private operators and not by the canal authorities. Towing by animals was banned as of 1908, except in the case of timber rafts that floated down the old locks, but an end was finally put to this exception in 1924. For shipowners, towing considerably slowed the progress of vessels along this waterway. For canal engineers, towing was detrimental to the condition of canal banks, since, as is illustrated in the photo, the canal was not yet delimited by stone or concrete walls along its entire length. In the Rockfield sector, which was so named because bedrock had to be excavated at this location, the shores were laid out as embankments. Several years went by before stone and concrete work was done to contain this stretch of the canal.

Wm. Notman & Son, *S.S.* Bonavista, *Montreal, Quebec*, McCord Museum, Notman Photographic Archives, II-106286, 1894

DOCKING REGULATIONS

Beginning in 1833, the canal commissioners stipulated that goods could not be left on docks for more than 48 hours. Seven years later, new regulations were applied to fuel wood, which from that time on, was not allowed to remain on docks for more than 24 hours. By the end of the 19th century, shipping traffic had considerably increased, and, for the first time, docking rights became subject to strict regulation. Henceforth, each shipping company was assigned a pier for the navigation season, with priority being accorded vessels that had come to unload, rather than load, their cargo. In the case of a ship without an assigned pier, it could unload in front of the pier belonging to another company. In 1908, docking priority was granted to pier leaseholders. However, upper time limits were established for loading and unloading: 24 hours for loading a ship with 50 tons or less, 48 hours for 100 tons, with an extra day being granted for each additional 50 tons. Unloading was to proceed at a rate of 75 tons per day and was not allowed to obstruct movement on docks. Freight for loading was not to remain on docks for more than 36 hours. In the case of ships supplying produce markets, the parent companies were entitled to erect a warehouse on the mooring wharf. However, docking fees were not applied to goods that were unloaded directly into warehouses or on the property of industrial plants without transiting via the piers.

The Lachine Canal, Viewed from the Côte St. Paul Bridge, St. Henri Historical Society, 60-ph-4, ca 1900

RULES OF THE ROAD

Just as lock docking operations were governed by a system of right of way, crossing and passing on reaches were also regulated by rules of the road. From the time the canal was made open to navigation, it was prescribed that whenever two ships encountered one another, the ship descending with the current had to move over to the side of the canal next to the towing path, while the ship heading upstream had to yield the right of way. In the early 20th century, ship engines did not all have the same power, so when one ship overtook another, slower vessel, it first had to signal its presence and then pass if it was more than 90 metres from a lock. Beginning in 1924, crossing was to be performed on the *starboard side*,* with the exception of tugboats and other small ships, as is shown in this photo. Eight years later, the rules were changed: whenever a boat wanted to pass another, it had to whistle once to signal its intention of passing on the *port side*,* or twice if it wanted to pass on the starboard side. The captain of the ship ahead had to whistle back the same number of times, as a way of saying "roger," or blow the whistle five times if he thought that passing was too risky. As this code proved too troublesome, ships returned to the rule of passing on the port side in 1937. At long last, this seemingly unending cacophony died down—much to the relief of canal sector residents.

* *starboard side*: right side
* *port side*: left side

Barge Stuck on Lock Sill, National Archives of Canada, Andrew Merrilees collection, PA 198567, undated

INCIDENTS ON THE CANAL

For a canal administrator, a marine incident was viewed in terms of how long navigation was interrupted, whereas for a shipowner, the same term was equated with damage. Between 1845 and 1945, the waterway administrators recorded some 225 incidents responsible for halting navigation. Whether these mishaps involved collisions with locks, as is illustrated by this barge accident, or collisions with bridges, shipping traffic could be closed for periods lasting anywhere from several hours to five days, as was the case with the incident involving the *Bohemian* in 1880. For the years 1917 to 1944 alone, the directors of Canada Steamship Lines recorded more than 160 incidents involving their fleet, including the ships of the company's subsidiaries. Two times out of three, these incidents involved collisions either with canal structures or with other ships or barges. However, three times out of ten, the ship had run aground or damaged its propeller. This type of accident seemed to correspond to the low water levels occasionally registered for the canal, as in 1931; indeed, most of these incidents occurred between late August and the end of the navigation season, when water levels were lowest.

HAYWARD, S. J., *S.S.* Barrie *of Canada Steamship Lines*, National Archives of Canada, Andrew Merrilees collection, PA 202664, undated

THE DURATION OF A JOURNEY FROM MONTREAL TO THE GREAT LAKES

In the earliest days of the canal, it took a batteau or a Durham barge 12 days to complete the one-way trip between Montreal and Kingston carrying a cargo of no more than 8 tons, on account of rapids and towing. The return trip went slightly faster, thanks to the current and the prevailing winds. Loads headed for Montreal were considerably heavier, attaining up to 40 tons in the case of flour, for example. A century later, travel time had become significantly shorter. In 1935, a bulk carrier of Canada Steamship Lines could go to Chicago with a load of corn and come back to Montreal with grain for the Ogilvie Flour Mills in exactly 11 3/4 days roundtrip, a figure that included the time required to load and unload freight. For comparison's sake, it is worth noting that Chicago lies 2,000 kilometres inland from Montreal, but Kingston only 280 (the distance had thus increased sevenfold), and that the freight carried no longer weighed 40 tons but instead 2,500 tons (that is, 60 times more cargo than in 1825).

Shipowners and Shipping Lines

Canal Near Dominion Wire Works, Looking West,
National Archives of Canada, Andrew Merrilees collection, PA 202565, 1903

THE MONTREAL TRANSPORTATION COMPANY

From 1825 to 1945, the ships of approximately 50 shipping companies travelled up and down the canal. One of these companies, the Montreal Transportation Company, is particularly noteworthy. As indicated by its corporate name, this shipping company was based in Montreal. One of its most important shareholders was Hugh McLennan, the leaseholder of one of the hydraulic lots along the second reach and the owner of a grain elevator. McLennan and his partners (who included J.-B. Auger), bought up the fleet of Chaffey Bros. of Kingston, which had gone bankrupt. George Chaffey would nevertheless be brought into the firm as a partner. Their goal was to become the most powerful inland navigation company in Canada. McLennan later became president of the Montreal Warehousing Company, which was also established along the canal next door to the flour basins and which specialized in handling grain and flour. The Montreal Transportation Company board of directors included among its members William Watson Ogilvie, the owner of the Ogilvie Flour Mills, the largest flour mill operator in the Commonwealth. In 1892, the company owned 29 barges for use on the St. Lawrence, six for use on the Great Lakes, and nine tugboats. During this decade alone, it carried two-thirds of the grain shipped to Montreal by boat from Kingston. Thirty years later, on the eve of its acquisition by Canada Steamship Lines, the fleet numbered 12 steamships, six barges, and four tugboats.

Steamer Ship Beauharnois *in St. Gabriel Lock, Lachine Canal*, National Archives of Canada, Department of Public Works collection, PA 183231, 1910

ABOUT THE 19TH-CENTURY ANCESTOR, THE SOCIÉTÉ DE NAVIGATION DU RICHELIEU...

The Société de Navigation du Richelieu was started up in October 1845, at the same time that it came into ownership of the *Richelieu*, a steamship built by William Parkyn, who later became a leaseholder of Côte St. Paul hydraulic lots. It was the beginning of a long adventure that has continued until our time. Thirty-five shareholders, including Jacques-Félix Sincennes, a merchant from Sorel, Quebec, and Victor Hudon, a future textiles businessman in Hochelaga, Quebec, decided to operate a shipping company that would carry passengers and first-class freight, in addition to providing a towing service between Montreal and Lake Champlain. Three years after its founding, the corporation bought out a rival and became the Compagnie du Richelieu: this was the first in a long series of acquisitions and trades. In 1853, as operations in the Richelieu valley proved unprofitable, the company withdrew from the area and fell back on Montreal and the St. Lawrence. After appointing David Torrance, a leading shipper on the lower St. Lawrence, to its board of directors, the company signed an agreement with the Royal Main Line, which covered routes along the upper St. Lawrence, with the objective of reducing competition. The Royal subsequently encountered financial hardship, which in turn enabled the Canadian Navigation Company and Hugh Allan, one of the most prosperous of the Montreal shipowners and a specialist in trans-Atlantic transportation, to take over this company. The fortunes of Canadian Navigation and the Compagnie du Richelieu were officially united in 1875. Now, the Richelieu and Ontario Navigation Company served the entire St. Lawrence River from Toronto to the Saguenay River area. Upon the conclusion of this merger, it had 18 steamships specializing in passenger transportation. Included among these ships was the *Richelieu*, which was rechristened the *Beauharnois* in 1906. That year, the *Beauharnois* was the oldest steamer in operation in Canada; at 40 metres in length, 8 metres across the beam and 2.1 metres in depth, the steamer's dimensions gave away its venerable age.

Kitts R., *The Arrival at Lachine of* S.S. Selkirk, *Loaded with Hot Steel from the Stelco Plant, Which Resulted in a Two-Week Stoppage at Stelco's Dominion Works*, National Archives of Canada, United Steelworkers of America collection, PA 120661, 1946

… AND ITS MODERN-DAY DESCENDANT, CANADA STEAMSHIP LINES

On the eve of World War I, the Canadian merchant marine was in a state of flux. It was about to witness a succession of mergers, one of which produced Canada Steamship Lines. At the time of its creation, the company operated more than 103 ships on the Great Lakes and held a near monopoly over river traffic. Among the 10 companies forming the new consortium was the Richelieu and Ontario Navigation Company. It would be joined in 1920 by the Montreal Transportation Company. Behind the scenes, major stakeholders in the Lachine Canal had a hand in these mergers. People like Ogilvie and McLennan are already well-known figures, while other actors are considerably less so. The Richelieu and Ontario shared close links with the Grand Trunk, whereas Inland Navigation had ties to the new Steel Company of Canada, which owned four factories along the canal. In fact, the new company brought together shareholders not only from Montreal but also from Ontario and Manitoba, thus providing a good illustration of the Canadian maritime "network." British interests were also prominent, as Canada Steamship Lines had an advisory board that included, notably, a representative of the Furness Withy company, one of the largest shipowners of Great Britain, and another from Albert Vickers, the famous shipbuilder. Despite getting off to a shaky start, Canada Steamship Lines stabilized its financial situation during the war years. The company was able, in 1926, to acquire the Canadian assets of the George Hall Corporation, based in Ogdensburg, New York, and of Hall Coal, based in Montreal. In the late 1920s, it took over Century Coal, another in a string of assets located up and down the Lachine Canal.

Shipbuilding and Repairing

HENDERSON, Alexander, *River Boats and Activities at the Harbour*, National Archives of Canada, John A. Brooke collection, C 004850, ca 1875

SHIPBUILDING DURING THE ERA OF THE CANTIN SHIPYARDS

Augustin Cantin's Canada Marine Works was the leading shipyard on the Lachine Canal and one of the largest in Montreal during the 19th century. Cantin, a native of Cap-Santé in the region of Québec, produced alone or in a partnership with E. E. Gilbert more than 100 ships, all of them wood constructions. His most important clients included John Torrance, the Grand Trunk Railway, the Ottawa River Navigation Company, the Gildersleeves of Kingston, and the Allan Lines. Also figuring on this rather impressive list was Jean-Félix Sincennes of the Société de Navigation du Richelieu, for whom in 1855 he built the *Cultivateur*, a steam side paddle-wheeler measuring 47 by 7 metres. Until 1892, Cantin produced either side paddle wheelers or screw-propelled steamers. It is worth noting that, beginning in 1875, Cantin produced more tugboats than major ships. This shift in direction was commanded by the emergence of new competitors, first in Ontario and then in Great Britain. Times had changed in the shipbuilding sector, and as competition grew increasingly fierce, shipping companies now did business with Ontario shipyards, whether in Toronto, Kingston, or Collingwood. On the other hand, these same shipyards, along with those in Montreal, were hard put to share in the expansion of shipping that occurred during the early 20th century, when shipowners bought steel-hulled ships of British make which, besides being adapted to the locks of the waterway, also cost less than Canadian-built vessels.

Launching of New Tug Lachine Canal,
Parks Canada, Quebec Canals, 1931

THE ANCESTOR OF THE MONTREAL DRY DOCKS

When brothers George and William Tate leased the land for setting up their dry dock in 1851, they were by no means newcomers to the Lachine Canal. They had already leased the three former locks of the first canal in Montreal that they had converted into refitting yards. A sawmill and a nail works were also located on the site of their dry dock, next to the shipbuilding yard, and were run using the hydraulic power supplied by the canal. Although the Tate brothers are best known for refitting, they also built ships, in particular, the *Louis Renaud*, a 52-metre ship that ran aground in Lake St. Louis in 1873. The Tate Brothers shipyard was bought out in 1895-1896 by the Lachine Canal administration, which was particularly interested in finding a wintering location for the flotilla of canal dredgers. On the eve of World War I, the property was taken over by Hall Engineering, which set up the Montreal Dry Docks on this site. The new administration added a second dry dock and, most likely, a cradle used to launch ships. Bought out by Canadian Vickers in 1928, the shipyard continued to operate until the canal was finally closed.

Thériault, G., *Rhine River Barges, Hulls 56, 57 & 58 Loaded on Scows Ready for Shipment*, National Archives of Canada, Dominion Bridge collection, PA 202853, 1947

LITTLE-KNOWN SHIPYARDS

A quick survey turns up seven shipbuilding yards that settled along the Lachine Canal between 1848 and 1945. The largest of these shipyards was that of Augustin Cantin, which went by the name of the Canada Marine Works. The others were the Tate Brothers' dry dock in the vicinity of Mill Street; the docks of Augustin Labbé located on the former island facing the Cantin yard; the boat works of Jean-Baptiste Auger a slight ways upstream from the previous two outfits; and the shipyard of the Allan brothers located in the Brewster basin near present-day Atwater Street. Auger later became a shareholder in the Montreal Transportation Company. During World War I, a shipyard was established and operated by Fraser, Brace and Co. of Trois-Rivières, Quebec, on the south shore of the canal, a slight distance upstream from the others. However, two years later, this shipyard closed down. Its facilities were recovered by Canadian Power Boat, which produced speedboats for the Royal Canadian Navy beginning in 1941; this company's activities came to an end with the conclusion of World War II. Something of an oddity among the Lachine Canal shipyards was the Dominion Bridge dockyard, where, for the duration of a few contracts, Rhine River barges were turned out at end of World War II. This venture was not the product of chance, since even back during the World War I, Dominion Bridge had also designed ships and barges.

Bulk Transport

Elevator B Explosion, Port of Montreal Archives, photo 3192, 1924

WHEAT AND FLOUR

One of the goals of the promoters of the Lachine Canal project was to carry raw materials produced in the newly settled areas of Ontario to Montreal for reshipment to foreign markets. Wheat and flour, along with wood (and, beginning in 1850, coal), were the main commodities transiting through the Lachine Canal. In fact, these products accounted for an average of 60% of the total carried over the canal. Although some years, such as 1845-1848, wheat and flour represented more than 75% of the freight bound for Montreal from the Great Lakes, as a rule, this proportion hovered in the neighbourhood of 25%. Until 1860, Ontario shipped more flour than grain, but from that year on, the reverse situation prevailed, as wheat became by far the leading staple delivered in Montreal. This era also saw the emergence of grain warehouses, most of them concentrated in the Montreal harbour sector, although Hugh McLennan established an elevator on reach no. 2 around 1850. Elevator B (referred to today as Silo no. 5) was also used to store grain. Built in 1905 by the American firm John S. Metcalf, whose founder was a native son of Montreal, this elevator is the only one to have a structure made entirely of steel. Above the elevator is the conveyer section, which rises a further six floors. The annex on the right was also built according to the Metcalf plans. Erected in two phases, in 1913 and 1923, the second elevator consisted of 95 cells, and was built entirely in reinforced concrete so as to offer better protection in the event of an explosion. The last elevator in this complex (and the longest, although absent from this photo) was constructed in 1958.

Fred Carney *at Coal Docks*, McCord Museum, Notman Photographic Archives, 106957-II, 1894

COAL MOVEMENTS

Coal entered the holds of bulk shippers only beginning in 1850. Coal accounted for an average of 20% of products transiting on the canal, although it represented the number-one commodity during both world wars. Contrary to wheat and flour, however, coal was shipped in both directions over the canal system. Until around 1875, it was Nova Scotian and, to an even greater extent, British coal wholesalers who exported their coal to Ontario. Beginning at this time, and until the onset of the Great Depression in 1929 (i.e., for close to 60 years), Pennsylvanian coal ranked foremost among shipments of this resource. Nova Scotian coal turned the hard times to good account, taking back first place, but its fortunes again shifted with the outbreak of World War II, when American coal regained the ground previously lost. The cross-hauling of coal, at least during the 20th century, partially reflected the Canadian government's coal provisioning policy, which attempted to preserve the mining interests of Nova Scotia by subsidizing coal transportation from that province. Appearing in this photo is the *Fred Carney*, a coal steamer belonging to the Ogdensburg Coal and Towing Company, which delivered coal from Pennsylvania and Ohio to Montreal.

St. Gabriel Lock, Lachine Canal, National Archives of Canada, Andrew Merrilees collection, PA 198570, ca 1925

LUMBER AND PULP

Wood taken from the new lands being cleared in Ontario constituted the leading product to transit via the canal, at least during its first decades. Until the early 1870s, lumber —whether in the form of squared timber, firewood, staves, spars or shingles—accounted for close to 45% of the products carried on the canal in years both good and bad, and would represent an even greater share if by-products such as potash were included. From the late 19th century, the proportion of lumber in waterway freight totals continually declined. Beginning in 1907, however, wood movement picked up again—this time in the opposite direction, since pulpwood was now loaded onto barges headed toward the Great Lakes for use in manufacturing newsprint. Nevertheless, the extent of pulpwood movement was limited, for wood accounted for scarcely 16% of the combined total for lumber and pulp cargoes between 1907 and 1920, and this proportion fell to 10% between 1926 and 1946. This trend was clearly the reflection of several political actions, including, on the one hand, the application of an American tariff on pulp imports and, on the other, the enactment of an Order in Council of the Quebec government in 1910 banning the export of pulpwood harvested on Crown lands, with the result that only lumber from private woodlots could be exported. Furthermore, beginning in 1920, a substantial increase in stumpage fees and a conservation policy for public forests were adopted, contributing significantly to this drop-off. All the same, between 1925 and 1933, the *Chicago Tribune* had its designated pier on the St. Gabriel basin, where it loaded its pulpwood. Although the paper industry grew to its position of pre-eminence in the Quebec economy during the interwar period, statistics on paper shipments only appeared beginning in 1949.

Lachine Canal, circa 1900,
St. Henri Historical Society,
60-ph-5, ca 1900

MONTREAL-BOUND

This photo of a ship travelling up the river would appear to stand in contradiction with the title. But this seemingly innocuous photo nevertheless portrays a reality. While only 12,700 tons of freight transited through the Lachine Canal in 1825, the year of its opening, the movement continually intensified over the following decades. On the eve of World War II, transported tonnage totalled close to 9 million, making for an exponential increase. However, this movement of goods only really took off beginning in 1897. In fact, if year 1900 were assigned an indicator of 100, this figure would rise to 184 in 1909, 285 in 1921, 472 in 1930 and 687 in 1938. Prior to 1897, this indicator remained well below the base level. Nevertheless, this movement was practically only one-way, as load-bearing traffic tended to flow downriver throughout the entire history of canal shipping on the St. Lawrence. In 1825, only 400 of the 12,700 tons of merchandise recorded for that year travelled up the canal. During the following century, three times more freight was moved downstream than was moved upstream, as part of a pattern that was only altered over a three-year period, between 1934 and 1937, when refined petroleum from east Montreal was shipped to Toronto. After World War II, this proportion slackened somewhat, descending to 60%. As a corollary to this rule, ships sailing in the direction of the Great Lakes often travelled light. No better proof of this is to be found than the photo appearing here, which shows the scale used to measure the ship draught.

Loading Sacks of Grain on Ship, McCord Museum, Notman Photographic Archives, MP-0000.587.126, ca 1920

CONTENDER OR PARTNER?

When the Grand Trunk Railway established its switching yard and shops near the canal at Point St. Charles, not a few shipowners began to fear for the survival of their shipping companies. As they were unable to compete with trains during the winter, these firms pondered their ability to survive, especially if trains began to carry grain. Had they cast a glance Stateside, they might well have noticed that from 1869 to 1886, the proportion of grain carried by railway to New York City rose from 40% to 72%, at the expense of boats. Would the same situation be repeated in Canada, and above all in Montreal? From 1882 to 1886, the Grand Trunk Railway alone upped its market share by 13%—to 38% of the total. In 1891, this proportion had risen to 49%, despite a drop in toll fees in 1889 that brought cargo quantities carried by rail back down to their 1882 level. Thereafter, ship transport of grain fluctuated between 75 and 61%, depending on various fare reductions or on bumper crops, which severely taxed the hauling capacity of ships. A company such as Ogilvie eventually signed agreements with the Canadian Pacific Railway in order to send its wheat by train from the Prairies as far as Port Arthur, Ontario, and from there by boat to Montreal. In the end, the train supplemented the ship because it cost more to use in Canada.

Passenger Transportation

Valentine and Sons Publishing, *Immigrants Arriving at Montreal*, Bibliothèque nationale du Québec, Michel Bazinet collection, album 10-53-b, undated

GO WEST, YOUNG MAN

The waves of immigrants who were landed prior to Confederation left their mark on the Lachine Canal, since they supplied a portion of the manpower needed to build it. However, the migratory movement towards the new regions of the West following Confederation was no less intertwined with the history of the Lachine Canal. Although the Conservative government of John A. Macdonald founded its credo of nation-building on the occupation of Western territories, the economic crisis of 1874—whose fallout would be felt until the end of the 19th century—put a damper on the migratory phenomenon. In 1891, the territories of the Prairies had only 250,000 inhabitants at most. Not long afterwards, a veritable tidal wave of immigration unfurled over Canada that originated in any number of factors, including the recovery of the Western economies; the implementation of a new national immigration policy; the discovery of a new strain of wheat that met with increasing demand in Europe, which was then urbanizing at a stupendous rate; and, last but certainly not least, the presence of the Canadian Pacific railway beginning in 1885. Approximately three million people came to settle in Canada between 1896 and 1914, and a great many of this number headed westward. As a result of this influx, the provinces of Saskatchewan and Alberta were created in 1905. This mass movement had a direct impact on the number of passengers who transited over the St. Lawrence canals starting from Montreal. Passenger traffic surged by 75% between the 1890-1898 and 1906-1914 periods. The two world wars and the Great Depression would nevertheless bring these migratory movements to an end, with the result that immigration into the country fell off precipitously.

Lachine Wharf at 5 P.M., National Archives of Canada, Andrew Merrilees collection, PA 201809, 1907

TOURIST CIRCUITS

The transportation of passengers over rivers was certainly nothing new, because prior to the appearance of the airplane, it was the only means of intercontinental transportation. But one innovation that stands out rather prominently in the late 19th century was the all-inclusive sea- or river-faring tourist circuit. (It is true, however, that as of 1825, American travellers could avail themselves of "Fashionable Tours" starting from Lake Champlain and venturing as far as Montreal.) In 1895, the Richelieu and Ontario Navigation Company, which had specialized in passenger transportation from its inception, brought out its official guide describing the most spectacular locations along its St. Lawrence system route, from Niagara Falls to the Saguenay fjord. Thirty years later, this component of Canada Steamship Lines offered the wanderlust-struck a cruise from Duluth, Minnesota, to the Saguenay, and in its advertising extolled the luxury of its ships and the quality of its cuisine. This route was divided into sections, and the various charter vessels always navigated in the same waters, as their architecture was adapted to the corresponding bodies of water. Thus, three boats covered the Duluth–Windsor portion, six the Toronto–Montreal segment, and seven the Montreal–Saguenay corridor. In order to accommodate its clients, the company was also the proprietor of two luxury hotels, the Manoir Richelieu in La Malbaie, Quebec (Murray Bay), and the Hôtel Tadoussac in Tadoussac, Quebec. The *Rapids Prince*, built in Toronto in 1911, plied the Montreal–Toronto circuit. During the 1935 season, more than 50,000 people embarked for Montreal, while some 40,000 departed from Montreal for the lower St. Lawrence.

Canada Steamship Lines Steamship Rapids Prince *in Lachine Rapids near Montreal*, National Archives of Canada, Hayward Studios collection, PA 078547, undated

AWE AND ELATION

"And now we are before the fiercest, most celebrated, most difficult of navigation as well as the last of the great St. Lawrence rapids—the Lachine. A universal stillness reigns among the passengers on deck, and their hearts throb with a dubious expectation as they look forward to the glittering sheet of foaming breakers ahead, with their two little green islets, dashing through the spray. Human speech can find no tongue in such a scene, but awe and the overpowering sense of the mighty forces in raging activity around, inspires the thrilling stillness of a mingled fear and pleasure into every soul—fear at the awful possibility of some miscarriage in our descent, pleasure in the triumphant exhibition of the flash and cloud of the cascade, of the earthquake and foam-fire of the cataract combined with the howling multitude of waters and the vast sweep and surging of the ocean wave. In we plunge among the breakers, and the headlong current bears us towards the shelving and insidious rocks, sometimes hidden, sometimes disclosed to view with the dark suggestion of others couched unseen beneath the water. Deftly we pass them by within a few yards of their treacherous edges, through foam, through mountain billows, with our bows sometimes apparently submerged, through hurrying eddy and swirling whirlpool, through clouds of spray ascending from the churning abyss crowned with the iridescence of a hundred rainbows, and amid the thunderous voices of the surging deeps. A moment more, we have completed the descent and ride in tranquility the placid bosom of the river beneath, with a sense of relief born of the contemplated dangers past. Had we but deviated to right or left by so much as a few yards, or cast our length athwart the stream, we had been hurled, by the angry current, upon the rocks to utter wreck, or instantly capsized, submerged and rolled amid a raging wilderness of waves. But the cool hand and clear eye of the pilot is equal to the perilous work, and it is a notable fact that no accident of any consequence has ever happened, nor has a single life been lost in the course of many years of steamboat navigation on these wonderful rapids." How successful was the above description of the passage of the Lachine rapids, excerpted from a promotional brochure of Richelieu and Ontario Navigation, in attracting passengers with a craving for strong sensations? There can be no telling for sure. But there can be no denying either that excursions to the rapids were immensely popular.

Ottawa River Navigation Company Dock, Lachine Canal, National Archives of Canada, Andrew Merrilees collection, PA 198573, undated

FOAM AND FROTH

Whether it was a bulk carrier or a passenger ship, there was no getting around the need to supply the ship and crew, and passengers had to be fed, too, of course. Steamships had to be stocked with fuel wood or coal, and vessels of all kinds had to be painted and fitted out with rigging and lines. The Richelieu and Ontario Navigation is just one example of a shipping company that called on the services of Montreal's largest ships' stores suppliers, which were prominently established along the Lachine Canal. This being said, equipment represented only one facet of supply. The Richelieu carried passengers around the river system and even specialized in excursions. It offered ambience, emotion, pleasure. Onboard, passengers smoked Havanas or one of the cigars produced at Fortier's in Montreal, sipped Mumm champagne or absinthe to the lilt of an orchestra and, in an ultimate touch of refinement, savoured the cuisine of Victor Olivon, one of the most fashionable French restaurant owners in Montreal during this time. Upon arriving in Montreal, Olivon was hailed as a culinary virtuoso after concocting "one of those meals that mark an era and suffice to immortalize a chef." He prepared his cuisine using cuts of pork from the Laing Company, fresh fish from Larrivée et Raymond, and produce from Hudon et Hébert, one of the leading Montreal wholesalers.

Workforce and Infrastructures: Inloading Cargo

Lachine Canal, Weir No. 2, National Archives of Canada, Andrew Merrilees collection, PA 202357, 1916

"FOR DUST THOU ART, AND UNTO DUST SHALT THOU RETURN"

Since grain and flour represented the largest volumes of cargo unloaded in Montreal, except during the wars, it was only natural that floating cranes were used in the canal zone to transship grain from barges. This type of crane had existed since the 1870s and at least one Montreal company, the Montreal Grain Elevating Company, provided shipowners with such service. The company remained in business until World War I and perhaps for even longer. However, working conditions at this company were particularly arduous, as is shown in the following excerpt of testimony given by a *stevedore** during the proceedings of the 1887 Royal Commission on the Relations of Capital and Labor in Canada. According to John Brennan, "[When] a longshoreman is assigned to unload a cargo of wheat, [he] is exposed to an almost insufferable situation. He can spend up to five hours in the holds trimming the grain [i.e., directing the suction blower that moved grain to the warehouse], and yet not wear any mask or [protective] device to protect his lungs from wheat dust, aside from the occasional sponge or handkerchief knotted around the mouth." When asked by the chairman of the commission whether any appropriate equipment was available for preventing dust infiltration, Brennan replied that "Perhaps there was some, but [if so] it is not generally used here, and it is not sought for." When asked which of the two, wheat dust or coal dust, was the most harmful to human health, another stevedore, Robert McCoy, unhesitatingly declared that wheat dust was by far the most dangerous. Nevertheless, a grain handler was paid wages much lower than those of a coal handler. Impoverishment and unhealthy working conditions thus proved to be mutually reinforcing, and ultimately made grain handlers the unwitting authors of their own fall—materially speaking, at least.

* *stevedore*: individual or company who contracts to unload a ship.

Coal Derrick and the Keystone Coal Boat, Centre d'archives d'Hydro-Québec, Montreal Light, Heat and Power Consolidated collection (F9), LaSalle Gas, F9 / 700763, undated

COAL DUST MEMORIES

Coal ranked second among commodities in terms of the quantities that were transshipped at locations along the canal, except during the two world wars, when it came in first. Although, until World War I, all industrialists agreed that coal was essential to producing steam, their enthusiasm noticeably cooled when the issue of storing this fuel source came up. It stands to reason that flour mill operators would manifest some opposition on account of coal dust. Instead, however, fears first arose among the lumber merchants and sawmill operators established in the vicinity of the St. Gabriel basins that had been reserved for inland navigation. According to them, coal dust discoloured lumber and diminished its value. Although this opposition sprang up in reaction to a particular set of circumstances, it nevertheless occurred at just the right time, since the government was preparing to dig two new basins and revamp the Wellington basin. Thus, a modus vivendi took shape under which space was set aside for coal at some distance from the large flour mills and lumber yards. This development led to the raising of coal derricks on the Wellington basin that were subsequently occupied by Dominion Coal, as well as derricks belonging to Century Coal, at the mouth of the canal, Ogdensburg Coal, in the St. Gabriel basin, and LaSalle Gas.

Boat Which Was the Scene of a Dockers' Strike,
Bibliothèque nationale du Québec, E.-Z.
Massicotte albums, 1903

LABOUR RELATIONS HARDBALL

Another scene of labour and union history was played out on the wharves. Against the backdrop of the hard times prevailing during the late 19th century, half of the strikes that opposed longshoremen and shipping or docking companies—that is, no less than 10% of all conflicts in Montreal—occurred between 1877 and 1895. All these disputes arose in connection with money. The use of strikebreakers became widespread and, while the first settlements won workers modest gains, the tone subsequently hardened and no advantages were won. At best, workers had to make due with taking back their jobs under the same conditions as before or else lose them to strikebreakers. Topping the list of the hardliners was the Allan Lines, the firm which most often went head to head with the interests of the members of the Ship Laborer's Society (a ship's hands union). After 1896, prosperity returned at last, although labour disputes were on the rise in Montreal (more than 650 strikes were recorded between 1900 and 1945). Nevertheless, all was relatively quiet on the waterfront, with the dockers calling only about 15 strikes during this time. Dockers did not appear to fall in with the trend observed among other labour groups, whose strikes becoming increasingly frequent and caused ever greater numbers of lost working days. As a rule, except for the conflicts of 1903, July and November 1911, 1922, and 1935, dockside disputes were short-lived or only affected small crews.

Triggs, Stanley, *Lachine Canal*, McCord Museum, Notman Photographic Archives, MP 2361-39, 1968

THE END OF SHIPPING ON THE CANAL

The opening of the St. Lawrence Seaway in 1959 did not signify an immediate end to the Lachine Canal. Obviously, the canal mouth located in Montreal harbour was shut to traffic, but boats continued to use the canal from the Lachine entrance, as is shown in this photo, and would continue to do so for another ten years. The Lachine Canal owed this lease on life to the Montreal Coke and Manufacturing Company (popularly known as LaSalle Coke) and a few other manufacturing firms. LaSalle Coke had signed a firm 10-year supply contract with Hall Transportation for its coal needs. The drop-off was no less dramatic for all that: loaded and unloaded cargo totals for the canal decreased by 90%, plunging from close to 12 million tons one year to 1.25 tons the year after. In fact, coal, which represented only 8% of all cargo in 1958, successively accounted for 28% in 1962 and 64% in 1968. However, these last statistics are deceptive, because beginning in 1958, coal transiting through the canal decreased by more than 400,000 tons annually. In 1962 and in the years following, coal tonnage averaged only 340,000 tons per year. LaSalle Coke would eventually adopt a new mode of supply —the train—which it relied on until shutting down in 1977, seven years after the canal had closed.

The Road from Hydraulic Power to Electricity—by Way of Smoky Valley

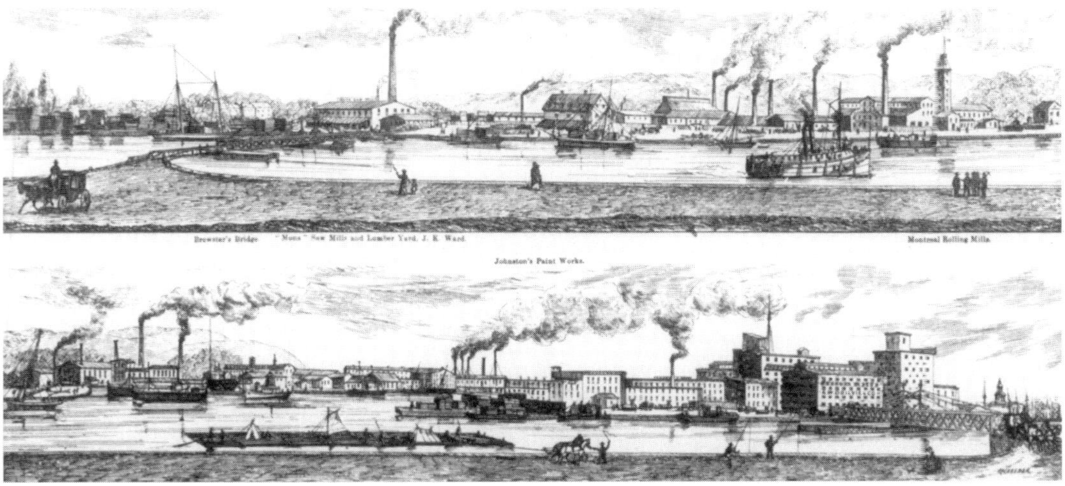

HABERER, Eugene, *Montreal: Panorama of the Lachine Canal,* National Archives of Canada, C 077224, July 1882

WHILE THE LACHINE CANAL is the cradle of the Canadian manufacturing industry, its rich history extends far beyond that sector's infancy. Indeed, what makes the canal deserving of more than a passing reference in a book is its unique position in our nation's history. The area has to its credit the dual feat of being not only one of the most impressive hydraulic complexes in the country, but also one of Canada's largest manufacturing complexes. Thus, from 1846 to 1945, more than 600 manufacturing firms from all sectors of production, including several of Canada's technological and corporate star performers, chose to settle there. Several factors were behind this success. First of all, water from the canal provided industrial establishments with driving force, as well as with cooling, bleaching, or cleaning agents. In addition, the canal became home to an industrial complex that resisted being confined to a single industrial technology, a single production management structure, a single production sector, or a single form of energy. Location was another factor in this success, since this site offered proximity to a commercial and financial metropolis, a pool of available manpower, a shipping canal, and a hub of transportation. Finally, this industrial area represented a critical mass from the viewpoint of companies for whom interdependence was an essential ingredient of their development.

THE RISE OF A HYDRAULIC COMPLEX

During the 1830s, Montreal's manufacturing and artisan-based production sector prospered, despite being only slightly mechanized. At that time, only hydraulic energy and steam were able to meet the combined criteria of power, accessibility, and low cost, but both had shortcomings which hampered the growth of mechanization in Montreal. For one, the city sorely lacked hydraulic sites, since those owned by the Séminaire de Saint-Sulpice had been taken over by nail makers and flour millers. Furthermore, to cover the costs of acquiring and maintaining a stationary steam engine, a manufacturer had to fully mechanize his operations; otherwise, operating such machinery was not economically viable.

By deciding to widen the Lachine Canal in 1841, the government of the United Canadas reconfigured the local energy profile. The considerable difference in ground levels along the new canal lent itself to creating a hydraulic complex similar to those of New England. In 1844, the Public Works Commissioners started up a pilot project on reach no. 2, their most promising site. In 1846, the first water privileges were leased. Five years later, this waterside location already presented one of the salient characteristics of the Lachine Canal manufacturing complex—that is, the prominence of companies from the iron and steel and the food and beverage sectors. The great diversity that became another particularity of the canal's industries was heralded by the presence of one firm each from the chemical, wood products, and transportation equipment sectors.

Encouraged by the success of the hydraulic pilot project, the Department of Public Works then sought to lease two more sites along the Lachine Canal that presented an energy potential—that is, the St. Gabriel locks and the Côte St. Paul locks. In contrast with the initiative at reach no. 2, in which the canal administration assumed the cost of laying out hydraulic facilities up to the industrial establishment itself, the future leaseholders of both sites had to foot the better part of the bill for developing the hydraulic lots. Even so, the St. Gabriel site held out enough potential that John Young and Ira Gould leased water privileges in 1851. When confronted with the scale of investments required to make the operation profitable, they subdivided their site into 17 different lots. Under the management of the St. Gabriel Hydraulic Company, these lots were quickly sublet. Given this subdivision, the St. Gabriel site not surprisingly attracted companies of a smaller scale than those located along reach no. 2. On the other hand, the St. Gabriel site presented a more diversified profile than elsewhere. While the St. Gabriel lots were eagerly snapped up, the plans of William Parkyn, the sole holder of water privileges at Côte St. Paul since 1853, did not fare as well. Although Parkyn had proposed a turnkey project to manufacturers, only one flourmill, two edge tool works, one nail works, and one sawmill chose to set up their facilities in what then was a far-off outskirt of Montreal. All in all, the Lachine Canal hydraulic complex met the mechanization needs of manufacturers intent on taking their establishments from a craft-like mode of production on to an industrial one.

The Beginnings of Smoky Valley

Despite the merits of water-powered industrialization, the Lachine Canal energy offer remained confined to a limited area, and, at any rate, by 1856, the hydraulic lots of reach no. 2 and the St. Gabriel locks had reached their point of saturation. Manufacturers who were excluded from this hydraulic complex but were interested in mechanizing their operations could not look to water power for a solution to their problem. They had to find an alternative or else pass up the opportunity to augment their turnover in response to growing demand. Consequently, several small-scale manufacturers decided to turn to steam. They were all the more inclined to use this energy source because Montreal had a huge coal market, offering black coal at a price that compared advantageously to that, for example, in Ontario. In other words, industrialists took advantage of the capacity of the city's import/export traders to play British coal off against Nova Scotian or Pennsylvanian coal and thus lay in a cheap supply. As a result, Montreal's manufacturing sector began to mechanize in earnest, and grew by leaps and bounds thereafter. The steam era truly came into its own in Montreal when John Redpath inaugurated a sugar refinery in the St. Gabriel sector in 1854.

Considering the Lachine Canal as a whole, the establishment of the Canada Sugar Refinery, better known as "Redpath," symbolized a new phase in the history of this waterway, for it heralded the spread of industrial development outside of the recently formed hydraulic complex. The arrival of Redpath in St. Gabriel also had another, symbolic value, for it was the first in a long line of industrial establishments that chose this waterway not for its energy potential but for its prime access to water, a seriously underestimated resource in industrial history. From that time on, the Lachine Canal increasingly deserved to be referred to as an industrial corridor.

Further impetus in this direction came with the establishment of the Grand Trunk Railway shops in Point

St. Charles in 1854. This had the effect of attracting or stimulating numerous industries associated with the construction of rolling stock, such as nail works, paint and varnish factories, sawmills, and foundries. In turn, such interdependence among the various industries acted as a catalyst, attracting other industries interested in taking advantage of the opportunities offered by such a concentration of companies. Its critical mass alone made the Lachine Canal industrial sector a magnet for industrial firms interested in settling in Montreal—all the more so because it was exceedingly rare to encounter the diversity and scale possessed by the components of this manufacturing core. As well, canal companies stood out for their tendency to produce goods not only for the local or Canadian markets, but also for the export market—which explains why several of them numbered among the Canadian leaders in their respective sectors.

What is more, this industrial area presented a profile that was unique in industrial Canada, for the canal's shores were home to large-scale industrial establishments that were powered either by hydraulic or steam energy. In 1871, the water-powered complex consisted of approximately 40 companies employing 2,007 people. The steam-operated category comprised some 20 establishments along the canal's industrial corridor and employed 2,082. These statistics alone amply illustrate that hydraulic power had already become something of a spent commodity, whereas steam, whose potential was not confined to a limited area, was in the throes of expansion. So much so, actually, that more and more factory chimneys were belching black smoke into the sky over southwest Montreal Island. In fact, this black smoke was such a recognizable feature of the Lachine Canal that, even though the sector had once been known as "Little Lowell," its inhabitants now called it "Smoky Valley."

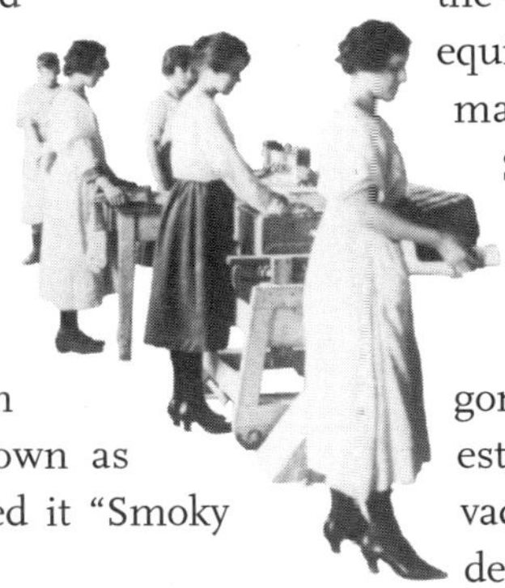

The Development of the Lachine Canal Industrial Corridor Continues Apace

During the last third of the 19th century, the Lachine Canal industrial corridor continued to grow vigorously; in particular, it benefited from the 1879 National Policy, the Canadian tariff policy which protected numerous enterprises from the dumping practised by other countries. The Montreal Rolling Mills took advantage of this trade barrier to diversify its production and began to manufacture iron tubing. The company did so well that, by 1881, it had become the leading Canadian producer. From 1880 to 1885, heavy industry grew stronger, with the establishment of Dominion Bridge, Dominion Barb Wire, and Dominion Wire Rope in the city of Lachine.

This tariff policy would also encourage members of the textile sector to settle along the banks of the canal. Thus, in 1881, the Merchants Manufacturing cotton mill set up in the St. Henri sector. In 1884, Belding Paul & Company, a silk manufacturer, built a plant near the St. Gabriel locks, while Dominion Wadding, which specialized in cotton padding and batting, was established near the Montreal Rolling Mills. Other types of industries also cropped up—for example, the Grand Trunk Export Stock Yards, which set up the first livestock market in 1882 at Point St. Charles, in an area where the slaughterhouse and meatpacking sector would eventually cluster. Similar developments include the expansion of the Royal Electric Company, an equipment manufacturer, and the arrival of the major paint companies, Canada Paint and Sherwin Williams. The industrial corridor of the canal had until that time been concentrated around three hydraulic sites and a few isolated locations in St. Henri, St. Cunégonde, and Point St. Charles, but now several establishments filled in a sizeable portion of its vacant areas and even created a new pole of development in Lachine.

In a separate development, while manufacturers of electric engines had entered the canal scene, electricity only hesitatingly penetrated factories in the area. Electricity was valued by companies not for its driving force but instead for its lighting applications. On that score, canal businesses were in the forefront, as the shops of the Grand Trunk Railway were the second in Montreal to use electricity for lighting purposes.

The Era of Large Companies

In 1896, Canada entered an era of prosperity founded on a combination of three factors: the expansion of the railway system, the wheat boom, and the upturn in industrial investment. However, the outstanding feature of these economic good times was a wave of horizontal concentration among establishments operating in the same branch of industry. In fact, during this period, regional companies that had previously shared the Canadian market among themselves now combined forces to achieve pan-Canadian status. Owing to the scale of its factories, the Lachine Canal would be at the heart of this trend. In the early 1900s, Montreal Rolling Mills bought out Hodgson Iron and Tube and Pillow-Hersey. In 1910, Montreal Rolling Mills and the Dominion Wire Manufacturing Company actively participated in the mega-merger that gave birth to Stelco. Two other sectors were to give rise to major amalgamations along the canal. In 1909, the Canada Car & Foundry Company of St. Henri and Dominion Car and Foundry of Ville Saint-Pierre merged with a Nova Scotian firm to form the Canadian Car and Foundry Company. In December 1910, this rolling stock firm acquired another major steelmaker, Canadian Steel Foundry, whose Montreal Steel Works located in Point St. Charles was its prize holding. In 1905, Merchants Manufacturing and the Colonial Bleaching and Printing Company amalgamated with other cotton mills to create Dominion Textile. In 1910, this giant of the cotton industry took over another cotton mill, the Mount Royal Spinning Company of Côte St. Paul (actually, it leased the Mount Royal facility until finally purchasing the unit in 1919, a few months before the lease expired).

This period was also witness to a wave of foreign, primarily American, investments in the Canadian manufacturing industry. In some cases, these investments resulted in the acquisition of Canadian enterprises, and in others in the establishment of new companies or branches in Canada. Owing to its density and scale, Lachine Canal industry saw the ownership of some companies transferred to foreign interests, as in the case of the Canada Paint Company or the Royal Electric Company. It also saw major American companies such as Sherwin Williams set up local operations. Furthermore, several new Canadian companies took advantage of this upswing to locate along the canal. In terms of industrial makeup, these new companies mostly belonged to the sectors of heavy industry, such as iron and steel or rolling stock; chemicals, such as manufactured gas; and electrical apparatus and supplies. With the arrival of these companies, most of the unoccupied areas of the Lachine Canal filled up. This was also the period when several companies gave up using hydraulic power and replaced it with steam, by then the dominant energy source in Montreal industry.

Even if World War I did not produce as strong a structuring impact as World War II, a number of industrial sectors nevertheless took advantage of the times to maximize their production facilities and even expand their physical plants. It was also during this period that electricity supplanted steam as a driving force in factories. The fall-off in demand for steam stemmed more from the combined effects of an upheaval in the coal market (constantly rising prices and increasingly difficult access) and subjective factors such as patriotic outpourings, than from a sudden attraction for hydroelectricity—particularly considering the tendency of industrial establishments to jealously preserve their energy self-sufficiency. Perhaps because of this, hydraulic power held up somewhat better in this new energy profile: even in 1940, it continued to be favoured by a group of die-hards, whereas steam had practically disappeared from the industrial landscape of southwest Montreal Island. In the meantime, however, the hydraulic complex had been withering away inexorably.

Following the mini-recession of 1920-1922, industry in Montreal entered headlong into a growth phase. Along the Lachine Canal, the city of LaSalle was selected by a number of chemical and petrochemical companies as a location for their operations and was witness to the transformation of LaSalle Gas into Montreal Coke and Manufacturing. Then, with the economic depression of the 1930s, each company had to do some close stocktaking and, above all, impose a hiatus on development, as was shown by the relative drop in investment spending by Montreal industry. All in all, however, canal industries followed the general trend of their respective production sectors. Thus, heavy industry and capital goods were harder hit than light industry and the consumer goods sector. However, this rule had its exceptions, as was shown by the good performances of two companies, Crane and Jenkins.

The salient feature of the Lachine Canal industrial corridor from 1860 to 1940 was its diversity. This fact is not altogether surprising, since the area always accounted for a third and at times probably half of the gross value of production of the Montreal Island industrial structure. Food and beverage industries represented an enormous proportion of the firms along the Lachine Canal. It could scarcely have been otherwise, considering that the canal banks were home to the main industrial establishments of the milling and meatpacking sectors, as well as "Redpath," one of the city's two sugar refineries. However, the predominance was not as overwhelming as it was elsewhere on the island. For one, heavy industry had a significant number of its largest production units in the Lachine Canal industrial corridor. This was particularly the case for companies in the iron and steel sector, which was represented by rolling mills, nail works, sheet metal shops, and iron and steel structure builders. In the rolling stock sector, the presence of the Grand Trunk Railway shops prompted the siting and growth of numerous related companies. However, it was not until the 20th century that two other rolling stock manufacturers—the Dominion Car and Foundry Company and the Canada Car & Foundry—set up permanent facilities along this industrial corridor.

Energy

Hydraulic Power

Reach No. 2, South Side, Parks Canada, Quebec Canals, 1910

A MAJOR-LEAGUE COMPLEX

During the 1820s, the Americans developed a huge hydraulic complex in Lowell, Massachusetts, to meet the needs of their energy-famished cotton mills. In Montreal at the time, any such initiative would have been useless, given the very modest needs of artisans and mill owners. Within 20 years, however, Montreal manufacturing had begun to take off, and the available supply of energy was poor, to say the least. To deal with the expense of widening the canal, the Public Works Commissioners viewed the leasing of water surpluses as an effective means of raising additional revenue. That is why, in 1844, they authorized the development of hydraulic lots on reach no. 2. However, an economic turndown slowed this pilot project, with the result that the site genuinely took off only in the 1850s, at the same time that the St. Gabriel and Côte St. Paul hydraulic sites were laid out. Then, the hydraulic complex truly hit its stride. During the following four decades, the 40 holders of hydraulic leases in this complex consumed great quantities of canal water in order to generate the hydraulic horsepower required to operate the industrial machinery of the nearby nail works, large-scale flour mills, sawmills and machine tool shops, wool and paint mills, to give but a few examples. This remarkable supply of power enabled the Lachine Canal complex to rise to a position of pre-eminence among North American hydraulic complexes while nevertheless remaining distinct on account of its diversified industrial structure. During the 20th century, the use of hydraulic power gradually died out on all of the three canal sites, although at a different pace in each.

Ogilvie Mills, St. Gabriel Lock, Lachine Canal, Looking East, National Archives of Canada, Department of Public Works collection, PA 183230, 1910

THE APPARENT AND NOT-SO-APPARENT SIDES OF A HYDRAULIC SYSTEM

It goes without saying that water was the indispensable mainstay of an entire system set up to convert this precious liquid into driving power. Fortunately for the users of this hydraulic complex, seasonal maintenance work on the canal and major construction projects sufficed to afford them *low-water periods** too. For without such reductions in water flow, the *headrace,** the *inlet opening,** and the *raceway** used to supply the waterwheels and hydraulic turbines of this industrial establishment would be drowned in a massive influx of water. Although this photo provides a glimpse of hydraulic infrastructures, it says nothing about the process by which water was converted into mechanical energy. Accordingly, the photo fits with the impression that passers-by would have had of hydraulic infrastructures as they strolled along the north side of the St. Gabriel locks. Only the workers in the industrial establishments concerned would have been able to see that the water routed through the inlet opening in fact served to turn a wheel or certain models of hydraulic turbines usually located below ground. Likewise, only employees were in a position to witness the conversion of this water into mechanical energy and its subsequent conveyance via a drive shaft and an assembly of gears, pulleys, and belts to one or more machines set out in the factory.

* *low-water period*: the period when the water level of a waterway is at its lowest.
* *headrace*: the reach of a channel or river that conveys water to a factory, weir, irrigation system, etc.
* *inlet opening*: the opening that allows the water used to supply a hydraulic machine to enter the industrial establishment.
**raceway*: the channel that directs water to the waterwheel.

Sin-Mac Tug Gertie *and Tow of Logs Passing Pillow & Hersey Mfg. Co.*, National Archives of Canada, Andrew Merrilees collection, PA 202882, undated. Actually, the photo depicts the Royal Flour Mills.

CATHEDRALS OF INDUSTRY

Owing to its strong difference in levels, reach no. 2 had the best hydraulic potential of the canal and was also located at Montreal's doorstep. With such assets, it comes as no surprise that the first hydraulic facilities (1845-1850) and the first industrial land uses (1847) were developed on this site. In five years, the 15 hydraulic lots available along the reach were leased out. The leaseholders built two large-scale flour mills, three nail works, one machine tool shop, one hydraulic grain elevator, one refitting basin (for ship repair and construction), one linseed oil and paint works, and one sawmill. Owing to the varied nature of their output, these establishments even then heralded the mix of heavy and light industry that would become a hallmark of the entire canal. And, with respect to their number, they prefigured the concentration of the site's hydraulic privileges in the hands of a few individuals or companies, as the flour mill operator Ira Gould, the machine tool producer W. P. Bartley, and *Grant, Hall and Co.** grabbed more than half of the available hydraulic power. During the following decades, two iron- and steelmakers, Peck and Benny, as well as Pillow, Hersey and Company, managed to take over most of the remaining lots. The industrial group of Bartley met with an opposite fate, as its lot was split into two distinct entities. This concentration led to the appearance of some of the most impressive 19th-century industrial buildings in the canal landscape.

* *Grant, Hall and Co.*: this sawmill, erected in the early 1850s, was converted into the Royal Flour Mills in 1859.

Lachine Canal, Lower Entrance to St. Gabriel Lock, Looking West, National Archives of Canada, Department of Public Works collection, PA 182060, 1910

HYDRAULICS—HOW PROFITABLE A VENTURE?

While this photo provides a good idea of the scale of works required to maintain a headrace, it is also revealing of the costs associated with such an operation. Interestingly enough, at the St. Gabriel locks, responsibility for this work did not fall to the Canadian government but instead to a private firm called the St. Gabriel Hydraulic and Land Company. As the holder of the hydraulic privileges of this exceptional site since 1852, the company had both to generate revenue and, at the same time, develop and maintain a number of hydraulic infrastructures. In order to make these operations profitable, the company's first four shareholders subdivided their five initial lots into 17 parcels. Furthermore, they allowed their leaseholders the liberty of subleasing surplus water to any takers. Thanks to this option, mechanization became a viable alternative for several small Montreal-based companies. Even if the major corporations were also content to sublease, the St. Gabriel hydraulic site primarily attracted smaller outfits than those located on reach no. 2. On the other hand, this site was considerably more diversified in terms of output. Thus, at various times, it was home to sawmills, flour mills, marble works, shipyards, and shops producing machine tools, edge tools, rope, textiles, paint, rubber, doors, and windows. Although in terms of use, the site proved an unqualified success, the bottom line at the St. Gabriel Hydraulic and Land Company was considerably less glorious, as upkeep ate away at its operating budget, reducing its income in the process.

Accident at Côte St. Paul Bridge, Looking East, Lachine Canal, National Archives of Canada, Department of Public Works collection, PA 110114, 1914

AN INDUSTRIAL ENCLAVE

Despite its excellent hydraulic potential, the Côte St. Paul site was spurned by industrialists because it lay at some remove from the urban core. In 1853, William Parkyn nevertheless took up the challenge of settling there, not so much in order to make use of hydraulic power as to repair and build boats. The laying of the new aqueduct canal nearby upset his plans for building a dry dock. But as he was bound by contract, Parkyn dug a feeder canal 600 metres long and 13.4 metres wide, thus endowing Montreal with a new hydraulic site, on which he built a 500-bushel capacity flour mill and several industrial buildings that he hoped to lease out. The response on the part of manufacturers did not match his own investment, since only one nail works, two edgetool works, and one sawmill set up on this location. In 1859, a battle-weary Parkyn transferred his rights to the Frothingham estate. Under its management, a new industrial complex was to emerge, featuring, notably, nail works, edgetool works, a sawmill, a cooper's shop, flour mills, and shops producing leather goods, bells, and cloth, not to mention the Mount Royal Rice Mills. During the 20th century, the urban and industrial fabric filled up the space between the St. Gabriel locks and the Côte St. Paul locks, thus putting an end to the isolation of this hydraulic site. Despite having lost its one-time autarky, the site nevertheless preserved a particular personality. Although with time, the companies on other locations progressively discontinued hydraulic power, a few of the Côte St. Paul factories formed a group of hold-outs that continued to use this type of energy until they were expropriated by the City of Montreal during the 1960s.

Lachine Canal, View Showing Entry to Dry Dock and General View of Weir No. 2 with Dredge Taking Away the Dam, National Archives of Canada, Department of Public Works collection, PA 151300, 1916

HE WHO LIVES BY THE MACHINE, DIES BY THE MACHINE

The water diverted toward waterwheels was the lifeblood of the machines of the Lachine Canal shops and works. Initially, however, this advanced mechanization occurred only in the industrial flour mills. Although nowadays they are commonplace, at the time of their introduction, automated flour mills inspired the boundless admiration of Montrealers. On the subject of the City Flour Mills, the French-language newspaper *La Minerve* wrote that, "There is no manual work involved from the moment the wheat is hauled into the boat until the moment when the barrel is ready to be loaded with fine wheaten flour, thus granting this mill a considerable advantage over less-developed mills."[1] This being said, the humming of machines continually intensified in the other industrial buildings of the hydraulic complex. Such an advanced degree of mechanization did not fail to surprise foreign visitors. Thus, for example, in 1864, Britisher Samuel Phillips Day frankly allowed that on the Lachine Canal, machines turned out products that were still made by hand in Europe. There was a dark cloud to this silver lining, however, since industrial growth ended up requiring increasingly more powerful machines and, ipso facto, a corresponding supply of energy. Although until the 1890s, hydraulic energy generally succeeded in assuaging this unslakable thirst, its use for industrial purposes had become equated with obsolescence in the iron-making sector some 30 years before. While hydraulic power remained a vital force, each new purchase of machinery further eroded its sway. One by one, the most outstanding representatives of hydraulic power use, such as the Royal Mills, gave it over for steam or electricity.

1. Description du procédé de fabrication de la farine au moulin d'Ira Gould, 1848, *La Minerve*, 05-11-1848

Coal and Steam

Barges Scotland *and* Isaac Stephenson *of the Ogdensburg Coal & Towing Co. in Company's Slip near McCord and William Streets, Lachine Canal, Montreal*, National Archives of Canada, Andrew Merrilees collection, PA 202637, 1863

THE WELLSPRING OF INDUSTRIAL ACTIVITY: FOSSIL COAL

Thanks to fossil coal, man changed night into day, winter into summer, iron ore into quality wrought iron, coal tar into a synthetic dye, and inanimate machines into powerful stationary and mobile steam engines. In fact, during the 19th century, the use of coal crept into all spheres of activity until it became indispensable to the functioning of society. In this new state of affairs, Montreal industry appeared to be cast in the loser's role, since no coal mine was to be found within a 1,000-km radius, and, transportation costs accounted for a significant portion of the price paid for coal. Fortunately, Montreal was able to turn this handicap into an asset by taking advantage of its position as a hub of the continental import-export trade. By playing the coal supplies of North America (Nova Scotia and the United States) off those of Europe (and especially those of Great Britain), Montrealers were able to purchase their coal at a competitive price. A catalyst of intense industrial and urban expansion on Montreal Island, coal also contributed to the strong growth of the major ship and rail transportation companies. In this respect, the Lachine Canal was particularly blessed, since as an industrial zone it also served as the focal point of ship and rail transportation in Montreal. Thus, it is not surprising that over the years a great many coal companies—i.e., coal vessels and wholesalers, including the Ogdensburg Coal & Towing Company in 1887—located their base of operations on the canal.

Canada Sugar Refinery Co. Ltd.,
Redpath Sugar Archives, 1897

WATER—A FACTOR OVERLOOKED IN THE TAKE-OFF OF SMOKY VALLEY

For a sugar refinery, water not only represented an essential "ingredient" (for dissolving, boiling, etc.) in the refining process, it also played a critical role as a thermal agent when it was converted into steam. By building his refinery in 1854 right next to the St. Gabriel locks, John Redpath thus pulled off a hat trick. First, he was ensured of constant, inexpensive access to water, a resource whose importance in the industrial economy has often been underestimated. Second, he solved a major logistical problem for himself by discharging his refinery's wastewater into the water supply source—that is, the canal. And third, he had docks on which to unload coal and raw sugar. With water available in abundance for its boilers and with builders of steam engines close by, the Canada Sugar Refinery Company (later known as Redpath) possessed all the assets required to generate steam for use as driving power on a scale unprecedented for the southwest sector of Montreal Island. Although the shops of the Grand Trunk railway took the technological lead in the late 1860s, the Redpath refinery's steam engines remained among the most powerful until the outfit converted to electricity during the between-war period. Indeed, the plume of smoke emerging from this factory hinted at the clouds of soot that were soon to spread over the future Smoky Valley.

Machine Shop, Canada Science and Technology Museum, Canadian National collection, 31125, 1929

AN INGENIOUS SET-UP

At first glance, this Canadian National Railway shop at Point St. Charles appears to be an inextricable tangle of metal gears and leather belts. But a closer look produces a radical change in perspective. The upper portion of the photo shows rows of metal shafts linking a group of pulleys that are used to transmit motive power to machines via a series of belts. Appearing prominently in the lower portion of the photo are the machines proper; upon closer inspection, they can be seen to be connected to the ceiling system by leather belts. In other words, the entire workshop is lined with shafting. Each drive shaft has a number of pulleys that convey the movement of the shaft to the machines. Owing to its size and complexity, this drive shaft system produced major, if underestimated, impact on industrial architecture. By so ingeniously and judiciously conveying driving power outward from a central point to ever increasing numbers of rapid, powerful machines, companies were able to reap considerable energy savings. This was a critical advantage, as a poorly designed transmission system could cause the needless loss of up to half of the energy produced for industrial uses.

Pinplat in Lachine Canal Berthed at Montreal Rolling Mills Co. Ltd., National Archives of Canada, Andrew Merrilees collection, PA 198572, ca 1910

ON THE CUTTING EDGE OF ENERGY USE

Depicted here is the Montreal Rolling Mills Company at the peak of its power, the very same year it joined up with three Ontarian companies and the Dominion Wire Manufacturing Company Limited to form the largest steelmaking firm in Canada, the Steel Company of Canada Limited, in Lachine. At the same time, this photo bears witness to a bygone era, when steam was used to drive the machinery of this huge steelmaking complex. A few years later, the Notre Dame St. workshop was equipped with electric furnaces, which produced better-quality steel than that turned out in open hearth furnaces, and was fitted out with electric rolling mills to produce rectangular steel ingots. These innovations were a sign of the times, since the same company had introduced the first steam rolling mill (1862) in Montreal, and, so doing, had made the hydraulic rolling mills of the nail works along reach no. 2 look like museum curios. Such upgrading was scarcely surprising on the part of a company that had extended its policy of remaining on the sector's cutting edge to its new entity, Stelco, even if this policy entailed substantial costs and across-the-board technological conversions. Stelco's wager paid off, as the company took first place in Canada for the production of steel ingots, while maintaining its position as industry leader for iron and steel processing.

Electricity

Côte St. Paul, Electric Station & Wooden Flume, Lachine Canal, Looking East, National Archives of Canada, Department of Transport collection, C 063600, 1905

HUNGERING AFTER POWER

In the dimly lit streets of gaslight-era Montreal, the first electric lamps burst upon the scene like miniature suns. To achieve such brilliant effect, electricity companies used small steam generators. These low-output power generating stations struggled mightily to light up the city's streets. The simple solution to the problem was to harness the current of a powerful waterway. Located a relatively short distance from the downtown area, the Lachine Canal quickly drew the covetous looks of promoters. However, the canal commissioners did not see things in the same light, and every overture by these companies ran up against their obstinate refusal. As the direct approach had proven ineffective, the companies then tried to obtain the rights to the hydraulic lots. There, too, they were turned down again and again until 1893, when the wind turned in the favour of Citizens Light and Power Company. This firm subleased a portion of the hydraulic privileges of the Frothingham estate (Côte St. Paul), and so doing, became the first Montreal electricity company to use the *kinetic** power of water, at least during the summer. In 1896, the canal administration followed in its footsteps, converting the former Tate Brothers sawmill located on reach no. 2 into an electrical power station and, most notably, building a small powerhouse on the Côte St. Paul locks in 1903-1904.

* *kinetic*: associated with or based on motion.

Raw Sugar Shed (Later Empty Lot next to Condé Street), Redpath Sugar Archives, ca 1904

AN ENLIGHTENING PERIOD...

Electric lighting—the theme of an international exposition in its time—made its 1879 Montreal début with the lighting of the Champ-de-Mars before thousands of astonished onlookers. Suddenly the faint glow of the gas lamp literally paled by comparison. Industrialists were quick to size up the import of this innovation. One year later, the furniture manufacturer J.-A. I. Craig and the Grand Trunk Railway shops in Point St. Charles became the first in the long line of companies that replaced candles and gas lamps with electric lighting. To accomplish this feat, they relied on a dynamo combined with a stationary steam engine. On the other hand, they did not venture so far as to supply their industrial equipment with any of the first electric engines. Admittedly, the time-tested technology of the stationary steam engine was still much more advantageous. The manufacturing sector was to remain loyal to steam throughout the 1890s, despite both the emergence of a new electric engine consisting of an alternator combined with a steam engine and the marked expansion of electricity companies. While, obviously, tramway companies no longer shied from using electricity as their driving power, manufacturers were long in coming round to applying this energy supply for other uses. Like a modern-day Trojan horse, electric lighting was admitted inside industrial establishments, including the Canada Sugar Refinery in 1893, thus laying the groundwork for the "electrifying" years.

Lower Entrance to North Lock No. 3, Gair Co. Ltd., Paper Board Mill (Looking North), Lachine Canal,
Parks Canada, Quebec Canals, 1942

... AND "ELECTRIFYING" TIMES

In Montreal, the electric years entered full swing during the late 19th century. There were signs aplenty that new times were at hand. For example, in 1898, the Royal Electric Company signed a sales contract under which it agreed to supply Dominion Cotton Mills with electricity for its driving power needs. Although this agreement did not touch off a rush for electric power, it nevertheless showed other industrial enterprises the way forward toward electrification. In spite of this, steam continued to dominate the city's industrial landscape, at least until World War I wrought its changes. Then, the pace of industrial electrification picked up, to the extent that by the end of the war, steam was already a fading memory for most industrialists. At the same time, Quebec experienced a wave of hydroelectricity-based industrialization of which the paper mills and aluminum mills are the best-known examples. While aluminum and pulp and paper were carving up kingdoms on the fringes of inhabited Quebec, no major aluminum or paper mill took up residence in Montreal. If it had not been for the contribution of the Lachine Canal industrial corridor, the regional portrait would have been much gloomier. For one, most of the electrical supply and device companies, which boasted fat backlogs, were installed in the sector. Furthermore, in 1916, Montreal Box and Paper Board set up a cardboard machine on the St. Gabriel locks. This plant is still in operation today, but under a different corporate name.

Industrial Star Performers

The Consumer Goods Industry

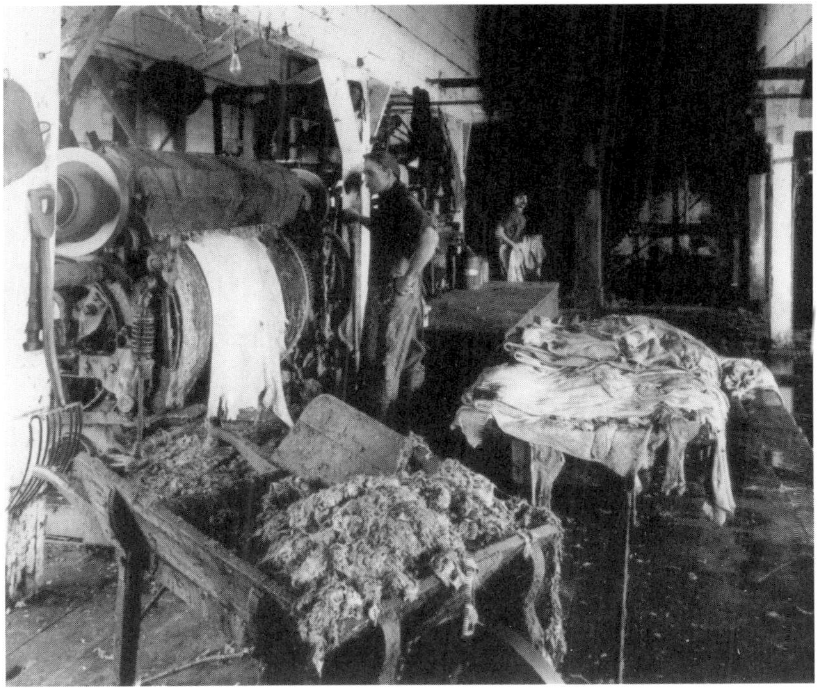

Keystone Pictures, *The Tannery around 1900*, Yves Beauregard collection, ca 1900

FROM TANNER TO LEATHER WORKER

A seasonal trade, tanning was done following the fall slaughter, when farmers brought large quantities of hides to tanners. An abundance of water was required for the repeated soakings and rinsings that turned these hides into leather. With brooks coursing down the nearby mountainside, the village of St. Henri was a centre for tanning in Montreal by 1685. Going by the name of St. Henri-des-Tanneries, and occasionally that of Tanneries des Rolland, this village in 1825 had 102 leather trade artisans, including approximately 30 shoemakers. In the 1850s, the Montreal shoe industry underwent mechanization, while tanners continued to work as artisans. Not surprisingly, they had to struggle in order to keep up with demand. The American leather merchant Edward V. Moseley leaped at this opportunity and, in 1859, inaugurated a tannery where hides for use in shoes, boots, and carriage leather were machine-tanned. The location along the banks of the canal in St. Henri assured Moseley of both a steady supply of water and a seasoned workforce. In 1871, Thomas Ecroyd also opened an industrial tannery. With its two tanneries, shoe works, and saddlery shops, the leather industry constituted a cornerstone of the St. Henri economy. Then, in 1882, the leather merchant A. Stevenson set his sights on this village-turned-city, building the Montreal Carriage and Leather Company on a lot bordering the Merchants Manufacturing Company. However, even at this time, the Montreal tanning industry had to compete against American rivals, which ultimately brought on its decline. After Moseley closed down his operations in 1911, only Montreal Carriage and Leather, a stone's throw from the canal, continued to operate, closing in the 1930s.

Meat Trolley for Mr. Scanlon, McCord Museum, Notman Photographic Archives, MP 112830-II,1895

LITTLE CHICAGO

The 1870s were boom years for live cargo exports to Great Britain. As the leading port for exports and the railway hub of the nation, Montreal became the unloading point for a steady influx of hogs and steers from Ontario and the West. Because it was a terminus for rail and shipping systems, the southwest end of Montreal in 1882 was chosen as the site of the Grand Trunk Export Stock Yards (Point St. Charles), the first in a series of livestock exchanges. Located in a brownfield, this stockyard eventually became the most complete facility of its kind in North America east of Chicago. Each year, thousands of head of livestock were herded off railcars and into pens, and from there to one of the neighbouring piers, from which they were herded back on to one of the trans-Atlantic ships heading for Europe, their final destination. A short time later, the Montreal Stock Yards, previously known as the Grand Trunk Stock Yards, supplemented its facilities with a slaughterhouse in which 5,000 steers, 5,000 hogs, and 5,000 sheep were killed each week. To achieve such results, the company had resolutely automated almost its entire production process. In a 1907 publication, the company even congratulated itself on its ability to process a live hog into a refrigerated carcass in less than five minutes. The sector took on the appearance of a little Chicago when the Laing Packing and Provision Company, a pork production and curing firm, and William Davies, a meatpacker, set up facilities there in 1901 and 1909, respectively. During the between-war period, in 1927, Laing (which had since become Canadian Packing) merged with Davies to form Canada Packers, the Canadian meatpacking giant. Montreal Stock Yards became Wilsil Limited in 1925, and remained independent until it was bought out by Canada Packers in 1955.

The Mount Royal Flour Mills, Borthwick, J.D., *Montreal, Its History, to Which ...*, Montreal, Drysdale & Co., 1875

THE COMPLEX PROCESS OF PRODUCING FLOUR

The seemingly simple process of reducing wheat into flour was in fact exceedingly complex—so complex that the experts readily refer to it in terms of a "system" and have developed sophisticated diagrams with which to illustrate it. While perhaps a bit oversimplified to some readers' liking, Reverend John Douglas Borthwick's description of the process has the advantage of going straight to the point. In the foreground of this photo, a vessel is moored near the Mount Royal Mills steam elevator. This structure had conveyer belts equipped with outward-facing buckets and was able to unload a ship's cargo at the rate of 3,000 bushels per day. Following storage in an enormous warehouse with a capacity for 750,000 bushels of wheat, grain was carefully screened so as to remove the slightest foreign substance that might have adhered to the berry. Then, grain was poured into cleaning machines to remove any speck of dust or smut. After being meticulously scoured, the grain was crushed until a powder was obtained. The last phase consisted in sending this powder through a *bolter** in order to separate different grades of flour and remove bran particles from the flour. Depending on the level of quality desired, grinding and bolting operations could be repeated several times.

* *bolter*: a kind of sifter consisting of rollers covered with pieces of silk (bolting cloth) of varying finenesses.

Glenora Mills, Bibliothèque nationale du Québec, E.-Z. Massicotte albums, undated

A FAMILY EMPIRE WITHOUT EQUAL

When John Watson debarked in Montreal in 1801 with only two grist stones in tow, it never occurred to him that his arrival would mark the birth of a family-owned flour-milling empire without equal throughout the Western world. En 1811, he formed a partnership with his nephew and future son-in-law Alexander Ogilvie. Upon Watson's death in 1819, the son inherited his father's shares in the Ogilvie-Watson mill. In 1848, Ira Gould created a stir in the Montreal flour mill sector when he established City Flour Mills on reach no. 2. To meet this challenge, Ogilvie and Watson, in a partnership with their brother-in-law, James Goudie, and with Ogilvie's son, Alexander Walker Ogilvie, built Glenora Mills on the St. Gabriel locks in 1852. This family enterprise proved a success and, in 1855, the elder shareholders turned over the reins to the new generation represented by John and A. Walker Ogilvie. In 1860, their brother William Watson Ogilvie also joined the board of directors of A. W. Ogilvie and Co. When confronted with the devastating dumping practised by American flour mill operators during the 1870s, Ogilvie proved to be a formidable opponent, capable of taking on the U.S. giant, Pillsbury, on equal terms. This success can be ascribed both to the strong capacity of its senior executives to anticipate technological developments and to a corporate policy of expansion toward the granaries represented by Ontario and the West. Tabling on their business relations with Canadian Pacific Railway, the Ogilvies in 1883 gained a monopoly over the silage and shipment of wheat from the West to the Great Lakes. During the same decade, the company bought up the Royal and City companies, both of which were located on the banks of the canal, and became a shareholder in Montreal Transportation. Although the future of the family empire now lay in the West, the Montreal mills nevertheless remained Ogilvie's flagship outfits. When William Watson Ogilvie died in 1900, control of the company slipped out of the family's hands. Following its incorporation in 1902, the Ogilvie Flour Mills Company ranked as the largest milling company in the British Empire.

General View of Yards, Office Dry Kilns, Box Shop & Planing Mills, James Shearer Co. Ltd., Montreal, P.Q., National Archives of Canada, Department of Defence collection, PA 024655, ca 1918

A BATTLE OF THE TITANS

During the 1900s, the agricultural revolution was in full swing in the Canadian West, while a series of crop failures beset the United States. The result was an explosion in Canada's wheat and flour exports. At the time, however, Montreal, which was the nation's main grain-handling port, had only one large flour mill, Ogilvie's. With such positive prospects before it, Montreal was witness to the emergence of two new mills—the Dominion Flour Mills (1912-1914) in St. Henri, and the St. Lawrence Flour Mills (1911), now known as Robin Hood Multifoods and seen here in the centre of the photo at its St. Cunégonde location. In these new flour mills, wheat was washed, ground, turned into powder, bolted and bagged. However much these mills resembled their forerunners, Rev. Borthwick would have been hard put to find his bearings in them. The grist stones had now been replaced by chilled cast iron cylinders, and the bolters had been supplanted by *plansifters*.* The list of technical innovations was very long indeed, but however spectacular they might have been, Borthwick no doubt would have been less astounded by them than by the roundabout routing of grain and flour up and down floors during various phases of production. In 1925, the Lake of the Woods Milling Company took over Dominion Mills. Thanks to the reputation of their company brands, the three flour mills located along the canal were able to confront the harsh economic winds of the 1930s without collapsing. In accordance with the dictates of commercial rivalry Ogilvie in 1914 launched a cookbook to promote its Ogilvie brand flour; to offset the impact of this initiative, Lake of the Woods Milling Company brought out its cookbook for Five Roses flour the following year. In 1954, Ogilvie acquired Five Roses.

* *plansifter*: this device consisted of a large number of superimposed sieves shaken by an eccentric.

Cutting (Left) and Packing (Centre), Paris Lumps Packaging, V1 Building, Redpath Sugar Archives, ca 1926

A CANADIAN SUGAR REFINERY

The Europeans arrived in North America bringing with them their taste for cane sugar. An imported product, cane sugar was not within the reach of all purses, and many a settler preferred maple sugar. Despite a venture in Nova Scotia (1818-1837), the Canadian sugar refining industry only genuinely took off in 1854, when the Canada Sugar Refinery, under the impetus of John Redpath, began operations along the Lachine Canal. In its impressive industrial building, sugar was first extracted by diffusing it in boiling water; the resulting juice was filtered for any non-sugar substances; then, it was heated in evaporators; and finally, the syrup was cooked until it crystallized. Once the crystallization process was completed, Redpath obtained a first run of white sugar, a second run sugar, and a non-crystallized syrup, molasses, which its code of ethics forbade from distilling into alcohol. For commercial uses, sugar was bagged, packaged, placed in barrels, or made into loafs. After getting off to a slow start, this sugar refinery left its stamp on Canada, but had to give way momentarily in 1876, when confronted with American dumping practices. Thanks to the National Policy (1879), Redpath sugar became a household name throughout the country. In 1902, the Redpath family overcame its moral misgivings and began to distil alcohol from molasses in a building adjacent to its sugar refinery. In 1930, powerful Canada Sugar Refining Co. (Redpath) merged its operations with those of the Ontario-based Dominion Sugar Co. Ltd., thus becoming the premier sugar refiner in the country. In the newly created Canada and Dominion Sugar Co. Ltd., the Montreal refinery continued to be the flagship outfit for some time yet.

DOMINION TEXTILE CO., LTD., *General View in Card Room— Draw Frames in Foreground, Fly Frames Beyond*, ca 1928, National Archives of Canada, Dominion Textile Co., Ltd. Annual Report for Year Ended March 31st, 1928 / Part of File H.S-4 Annual Reports, pt. 1 1905-1949, ca 1928

KING COTTON, AN EXACTING SOVEREIGN

Raw cotton would never have become turkey red cotton, calico, and other fabrics without the combined efforts of carders, spinners, weavers, bleachers, dyers, and printers. Given the scale of the tasks to be accomplished, each craft gave birth to a different industry, such as spinning mills or dye-works. Gradually, *carding*,* spinning, weaving, and printing operations were mechanized. Insight into the results of mechanization can be had from this photo of the carding shop and the first phases in spinning production (*draw frame processes** and *fly frame processes**) at the Mount Royal Spinning Company. Likewise, thanks to the chemical revolution, cotton mills managed to bleach and fix the colours of cotton fabrics with good results. Parallel to this mechanization and *chemicalization** of the industry, there was a new trend among cotton companies to bring traditionally disparate operations together under the same roof. On the Lachine Canal, the Merchants Manufacturing Company (St. Henri) and the Mount Royal Spinning Company (Côte St. Paul) were two such generalist companies whose activities embraced carding, spinning, weaving, bleaching, dyeing, and printing cotton fabric. For some time at least, their wager appeared to pay off, since both companies ranked among the largest of their kind in Quebec. However, foreign competition and lean profits eventually caught up with them. As of 1905, Merchants became one of the mainstays of the new Dominion Textile. As a component of this monopoly, it discontinued bleaching and printing operations. In 1910, the Mount Royal, which had been created in 1907 and which also joined forces with Dominion Textile, came in for the same medicine that had been administered to Merchants.

* *carding*: an operation consisting both in separating fibres and in eliminating short fibres and small imperfections remaining in cotton.
* *drawing*: the draw frame was used to eliminate any imperfections found in cotton thread and to produce a filament yarn that was stronger and of better appearance.
* *fly frame*: the purpose of this operation was to continue extending the sliver or the strand and to insert a slight amount of twist in thread in preparation for spinning.
* *chemicalization*: the incorporation of chemistry into the production process of industries.

Dominion Textile Co., Ltd., *"Design—Artist at Work on Cretonne Design,"* National Archives of Canada, Dominion Textile Co., Ltd. Annual Report for Year Ended March 31st, 1928 / Part of File H.S-4 Annual Reports, pt. 1 1905-1949, ca 1928

MAN AND THE MACHINE

In industrial history, cotton mills hold a special, quasi-mythical place. For one, this industry stands out for having spearheaded the introduction of *machines.** And, in a corollary development, it is associated with the supplanting of artisans, who took pride in their craft and their manual dexterity, by wage labourers, who were literally placed in the service of a machine. Caught in the midst of this technological revolution, workers in the British cotton industry sometimes resorted to rioting in order to defend their dignity, or to wrecking the machines that had generated their deplorable living conditions and condemned them to unemployment. In response to the ineluctable progression of mechanization, workers eventually changed tactics and made do with stopping the mechanical looms instead of destroying them. During the era when the cotton industry settled and developed along the Lachine Canal, this shift in attitudes had already been interiorized. For the first generation of workers, often newly arrived from the countryside, mechanized carding, spinning, and weaving were already a part of the natural order of things. In fact, these workers were already confronted with the *automation** of several mechanical operations. However, in this sector and others, the trend toward mechanization and automation varied depending on whether work was performed in a spinning mill or in a cotton bleaching company—although, ultimately, the outcome was the same. Some operations nevertheless continued to require a human touch, such as the design of patterns printed on fabrics. For no machine was a match for a designer when it came to charming the buyer with an elegant pattern in harmonious hues, or to pleasing the factory owner with a palette that could be printed cheaply on cotton cloth.

* *machine*: a mechanism, which, propelled by mere driving power, performs the movements making up a technical operation that had previously been carried out by one or more people.

* *automation:* a process which, under a set of given conditions, operates or is accomplished without the intervention of a human agent.

Dominion Textile Co., Ltd. *"Printing—Front View of Printing Machines,"* National Archives of Canada, Dominion Textile Co., Ltd. Annual Report for Year Ended March 31st, 1928 / Part of File H.S-4 Annual Reports, pt. 1 1905-1949, ca 1928

Arless, Sr., Richard, *Aerial Photo of Dominion Textile (Merchants Manufacturing)*, National Archives of Canada, PA 211412, ca 1945

THE MERCHANTS MANUFACTURING COMPANY

In keeping with the National Policy, the Canadian government slapped a heavy customs tariff on imported cotton, and Canadian cotton goods suddenly became much sought after. As Canadian mills were few in number, they thus reaped substantial profits. In 1880, one group of Montreal merchants seized this opportunity to found the Merchants Manufacturing Company, and, in 1881, proceeded to build a factory in St. Henri. In this establishment, cotton was spun and woven and, in addition, cotton cloth was bleached for use in shirt-making. In 1883, the cotton goods sector was hit with a major crisis of overproduction, which prompted mills, including Merchants Manufacturing, to form a cartel. The following year, the cartel collapsed and the St. Henri cotton mill shut down for 17 weeks. After flirting with disaster, the mill returned to profitability in 1887; five years later, it launched into a major program of plant expansion and diversification of its product line, including the printing of cloth. Unfortunately, however, competing companies imitated its example, thus triggering a new cycle of overproduction, falling profits, and workforce cutbacks. In 1898, Merchants Manufacturing struggled to keep up with demand for bleached cotton goods. To meet this demand, it doubled the floor area of its facilities and created a new set-up for the older machinery in order to produce continuously. In 1900, the Canadian government lowered the tariff on English cotton by a third, a measure which immediately cut into the profits of Merchants Manufacturing. The time seemed right for a merger of Canada's leading cotton mills. In 1905, after indulging in some lengthy fence-sitting, the shareholders of Merchants Cotton Company came to an agreement with four other major companies, including the Colonial Bleaching and Printing Company, to found Dominion Textile. Merchants Manufacturing was now one unit of this new giant but, with its 110,000 bobbins, it remained the second largest cotton mill in the country. The complex of buildings comprising Merchants Manufacturing covered both sides of St. Ambroise Street. Its main building, located on the edge of the canal, extends from the chimney appearing in the centre of the photo to the building bordering the railway.

South Abutment of Bridge No. 3, Looking East, Lachine Canal, National Archives of Canada, Department of Public Works collection, PA 110015, 1942

THE SILK MILLS — BACKSTAGE TO FASHION

It is one of humanity's secret paradoxes—and one of its outstanding feats—to have transformed the dried spittle of silkworms into a noble cloth. In Victorian-era Canada, where gaudy and pretentious trimmings were the *dernier cri*, skeins of raw silk thread were particularly valued for weaving rustly frills and flounces. Silk was imported into the country until the American Frank Paul arrived in Montreal in 1877 and began manufacturing this product. In view of the success surrounding this pilot project, the Belding Paul & Company built a four-storey mill accompanied by a two-storey dye-works on the Lachine Canal. In another demonstration of farsightedness, Frank Paul ensured that several American technicians were on hand from the time of the mill's inauguration in 1884. The following year, he pulled off a genuine coup by purchasing the machinery of the Corriveau Silk Mills in Iberville, Quebec, in order to produce ribbons. In 1889, the American firm Corticelli set up shop in Saint-Jean (sur-Richelieu), Quebec, to turn out the same line of products as Belding's. After engaging in mutually ruinous competition, the Canadian subsidiaries of both American companies merged in 1911—an opportunity Canadian capitalists used to take control of the new firm. Even though its range of thread products and trimming materials continued to form the cornerstone of Belding, Frank Paul quickly took advantage of changing fashions to diversify his product line. From 1885 to 1895, his mill wove practically endless bolts of silk goods for women keen on replacing their crinolines with suppler fabrics. In the 1920s, silk stockings were the rage, and it was not long before Belding-Corticelli Limited launched into this new market.

Heavy Industry

Point St. Charles Main Shop & Coach Yard,
National Archives of Canada, Canadian
National collection, PA 203036, undated

AN INDUSTRIAL CITY UNTO ITSELF

In 1864, the editors of the Canada Railway Advertising Company provided a highly informative narrative of tour of this huge industrial complex of the Grand Trunk. They began their account with the shops where cars were built and repaired, describing how work proceeded simultaneously on several different cars, while sleeping cars were built in an area nearby. They then moved on to the second floor, where they admired how the workmen performed cabinet and decorative work on first-class cars. Their tour then continued on to the machine shop, where they observed the precision boring of wheels and the insertion of axles into these wheels by hydraulic force. The next stop was the erecting (or repair) shop, which some of the employees referred to as the "hospital"; here, the visitors looked on as workmen performed surgical-like operations on each damaged "iron horse." They were fascinated above all by the immense hydraulic screw jacks used to raise locomotives from the ground and lavished no end of praise on them. On leaving these spectacular premises, they entered the turning and machine shop, where the Grand Trunk had assembled every imaginable type of machinery, from planers to hydraulic presses, for use in shaping iron to specifications. Although they were suitably impressed by the feats of the workmen producing steam boilers and of the blacksmiths who busily forged pieces with the aid of a pneumatic drill in and around two hearths and five blast furnaces, it was clear the tour no longer held the same interest for them. Although yesterday's technical feats have now been forgotten, the scale of the Grand Trunk shops at Point St. Charles continues to fire the imagination as strongly as before.

Grand Trunk Railway Erecting Shop, Point St. Charles,
National Archives of Canada, Andrew Merrilees
collection, PA 201824, 1860

Vulcan Iron Works, National Archives of Canada, Andrew Merrilees collection, PA 202873, 1860

STEAM—THE JUGGERNAUT BEHIND CAST IRON PRODUCTION

In Montreal, the fortunes of cast iron and steam were closely intertwined. In 1811, the use of steam for industrial purposes followed upon the efforts that the entrepreneur John Molson devoted to building engines for his boats. Caught up in the wake of steamship development, the two main local players —that is, the Eagle Foundry, with shops in the immediate vicinity of the canal, and the Bennet & Henderson Foundry—occasionally manufactured stationary steam engines. Beginning in the 1850s, the expansion of the manufacturing industry and the railway system fuelled demand for steam engines, hydraulic turbines, industrial machinery and transportation equipment. During the following decades, steam engine foundries and rail component companies entered a golden age; in that connection, the Lachine Canal, with a dozen of its companies active in this sector, benefited most strongly. Quality products were manufactured at the Eagle Foundry, Canada Boiler Works, Canada Engine Works, Caledonian Iron Works, St. Lawrence Engine Works, Vulcan Iron Works and J. R. Weir, and this output lent an unmistakable local colour not only to the steam engines used by Montreal industry but also to the various components employed in building ships and rolling stock in Montreal.

The Gazette, Blooming Mill, McCord Museum, Notman Photographic Archives, C069-C120-1534, ca 1935

FROM HYDRAULIC ROLLING MILLS TO ELECTRIC ROLLING MILLS

Until the mid-19th century, Montreal nail works purchased their nail plates in Great Britain, where this product was cheap. Suddenly, however, prices shot up. For nail makers, the solution was cut-and-dried: it meant rolling metal locally. The hitch was that no one had ever rolled iron in Montreal before. Not one to be put off easily, Mansfield Holland, on reach no. 2, was the first to roll out iron plates in 1859. That same year, another reach nail works, operated by Thomas Peck, set up a hydraulic *rolling mill*.* Both nail works used scrap iron. But as the supply of this iron was limited, prices began to climb. As men of action, both nail makers decided to circumvent the process by producing their own *pigs*.* Toward that end they built *puddling** furnaces. Since both nail works boasted rolling mills and *smelters*,* they began looking more and more like iron works. In 1862, Mansfield Holland, with the financial assistance of the iron and hardware import house Morland and Watson & Co., built a new, steam-operated rolling mill and a smelter. In order to augment their cash assets, the owners of this rolling mill turned their firm into a shareholding company, called the Montreal Rolling Mills Company, in 1868. Strengthened by the resulting influx of money, the outfit expanded its line of wrought iron products, and in 1881 commenced turning out iron tubing, becoming the leading Canadian producer in this sector. During the 1890s, the Montreal Rolling Mills Co. became one of the first Canadian businesses, if not the first, to substitute steel for iron in its products. When this company was integrated into the Steel Company of Canada Limited (Stelco), it became the Notre Dame Street shop and replaced its steam rolling facility with an electric rolling mill during the 1910s.

* *rolling mill*: a machine consisting of iron or steel rolls having parallel axes and between which metal (usually hot) is introduced for the purpose of being flattened.

* *pig iron*: an ingot of crude iron obtained by casting and cooling the iron in a sand mould or ingot mould.

* *puddling*: the purification of impure pig iron by agitating a molten bath in an oxidizing atmosphere—i.e., an air furnace or a puddling furnace.

* *smelter*: furnace for refining cast iron.

Motor Car for Toronto Transportation Commission, Montreal Works,
Canadian Railroad Historical Association Archives, Canada Car and Foundry collection, P 009, C-1381 lot 510, 1921

KING OF THE (URBAN) ROAD

During the 1880s, the electric tramway grew by leaps and bounds in North American cities, which placed a premium on rapidity. In Montreal, only the obstruction of the president of Montreal Gas, Jesse Joseph, prevented the city from electrifying its network prior to 1892. From that point on, however, Montreal transport-users swore only by the tram. This revolution in urban transportation served to put an industrial stamp on tramway manufacturing. Companies such as the Montreal-based N. & A. C. Larivière (1886-1903) took the lion's share of this market. In 1905, the American company Simplex Railway Appliance Company, established in St. Henri since 1902, erected a huge rolling stock manufacturing plant in Ville Saint-Pierre, and formed up a new company, the Dominion Steel Car Company, the following year. In this plant, Dominion produced the frame and steel bearings of tramways and rail cars, while the factory of the Montreal firm known as the Canada Car and Foundry Company, located in Turcot village (St. Henri), installed the wood superstructure. In 1909, Dominion was the instigator of a Canadian first, when it built a steel-frame car. That same year, both companies combined with the Nova Scotia firm Rhodes Curry to form the largest company of its kind in the country. As the popularity of the railway and tramway continued on unabated, Canadian Car and Foundry ended up employing thousands of Montreal workers. By the 1920s, some of the lustre attaching to the modernity of the train and the tramway had worn off, particularly in comparison with the car and bus. In 1927, in what surely amounts to a sign of the times, Canadian Car and Foundry delivered car bodies to General Motors and, two years later, engine blocks to this same American automobile manufacturer. As promising as such diversification might be, it was nevertheless unable to offset the dramatic drop-off in rail car and tramway orders during the following decade, although it turned out to be its best long-term hope.

Shop Pr. and Lt. Canadian National Railway,
National Archives of Canada, Andrew
Merrilees collection, PA 202552, ca 1935

AN INTERDEPENDENT WORLD

In our industrialized societies, where the manufactured product reigns supreme, most people are unaware that an industrial company's best customers are often one of its fellow firms. This situation was particularly obvious in Montreal and even more so along the Lachine Canal, where, for example, the shops of the Canadian National Railway (formerly the Grand Trunk Railway) purchased the products of other industries (iron, steam boilers, cotton, paint, varnish, manufactured gas, etc.) as well as manufactured goods, such as this spectacular overhead travelling crane capable of lifting a locomotive of several tons and purchased from Dominion Bridge. In turn, these same supplier industries ordered other production goods in order to fill the orders they received from transportation sector industries, and so on and so forth. This first, classic example is paralleled by the example of the cotton industry, in which thread produced in spinning mills such as Dominion Textile was used to make the insulated wrapping of telephone wires and cables or to manufacture pneumatic products and linoleum. However, cotton manufacturers also represented clients of choice for other industries. Thus, the acid and alkali sector sold to firms in the textile bleaching industry such bleaching agents as sulfuric acid or caustic soda, while chemical firms supplied the dyeing industry with the pigments it needed. Numerous other examples can be found, but suffice it to say that these business-to-business dealings represent a fundamental aspect of industrial development, since they give rise to commercial and technological poles through which growth can be spread throughout an entire economy. In short, interdependence fosters a characteristic relationship between propulsive industries and satellite industries, with the former directing progress and the latter adapting to it.

Canada Marine Works, A. Cantin, Montreal, McCord Museum, Notman Photographic Archives, MP 154 / 81 Mtl Harbour / 9529, before 1857

THE CANTIN SHIPYARDS

The shipwright Augustin Cantin set up his first shipyard at the entrance to the Lachine Canal and Montreal harbour in 1841. It was a particularly well-chosen location for building ships. But though orders flowed in, Cantin nevertheless went bankrupt in 1843. However, the man had considerable wherewithal and, in 1846, Cantin opened a new shipyard in St. Cunégonde. Laid out around a dry dock and a refitting basin, the shipyard included hangers and a forge. Cantin later added to these facilities with a second basin to the west, an engine foundry, and a sawmill. From that time on, he was able to build and repair all kinds of ships from start to finish. This formula proved to be a success since, from 1846 to 1855, the Montreal Marine Works built more than 70 boats, including seven steamships. This striking illustration of the Cantin shipyard prior to 1857 accurately depicts the bustling activity generated by 200 to 250 workmen whose efforts were entirely devoted to producing turnkey delivery of ships. Appearances can be deceiving nonetheless, because in 1859, Cantin again experienced financial hardship. It should be mentioned that shipowners of the time were confronted with serious competition from the railway sector. In 1861, inland navigation began to regain some of its previous vigour, as did Montreal shipbuilding, in an offshoot to this growth. The recovery went slowly, and the Cantin shipyard was only able to return to financial health in 1865. Augustin Cantin died in 1893, at the very time the first iron-hulled ships were being launched from Canadian marine works. With his son, Charles-Albert Cantin, at the helm, Canada Marine Works took due note of this new trend, but nevertheless made a specialty of repairing and building wooden ships only.

Canadian Pacific Railway Lachine Bridge—Unloading Steel from Ship Cont. 4137, National Archives of Canada, Dominion Bridge collection, PA 201561, 1911

A CANADIAN MULTINATIONAL

By enacting the National Policy in 1879, the Canadian government erected a tariff wall around the country. During the same period, the Canadian Pacific was engaged in building a transcontinental railway, which obviously necessitated erecting numerous metal structures and bridges. With American companies being excluded from the market for all intents and purposes, in 1879, a group of businessmen formed up Toronto Bridge so as to profit from this bonanza. Under the management of the American Job Abbott, the Toronto workshops operated at full capacity. Three years later, Dominion Bridge was founded, at the same time that it integrated Toronto Bridge into its structure. The following year, the company's directors embarked on a plan to build a factory in Lachine for manufacturing iron and steel bridge superstructures; once this ambitious project was completed, however, the plant's production capacity would exceed Canadian demand. But with their first 300 employees busily building bridges for the largest rail companies, their wager appeared to have paid off handsomely. The company and its employees thus have to their credit the rail bridge in St. John, New Brunswick, and the Canadian Pacific Bridge crossing over the St. Lawrence between Kanahwake and Lachine, to mention just two of their best-known accomplishments. In the 1890s, Dominion Bridge went into manufacturing steel structures for buildings. For many an architect or civil engineer, this firm became the reference for steel structures, and numerous bridges and factories along the Lachine Canal would bear this company's stamp. During the 1920s, Dominion Bridge bought out several of its Canadian competitors. Badly hurt by the Great Depression, the Lachine-based company laid off many of its thousands-strong workforce; only with the awarding of the Lion's Gate contract in Vancouver in 1937 did the company genuinely manage to recover. But recover it did—so much so that, 50 years later, Dominion Bridge had risen to multinational status.

Manufacture of Beam for Quebec Bridge, P.Q., Sun in Interior of Rockfield, P.Q. Plant of St. Lawrence Bridge Company, National Archives of Canada, Royal Commission on the Quebec Bridge collection, PA 108792, 1914

A CHALLENGE FOR CANADIAN CIVIL ENGINEERING

When Dominion Bridge was awarded the contract to build the Quebec Bridge superstructure in 1910, the company had arrived at a turning point in its history. The technical challenge facing this company was colossal, for it now had to build—over the St. Lawrence river near the city of Québec—the superstructure of a *cantilever** bridge having a central span of 548.6 metres and overhead space at high tide of 45.7 metres. But there was also a symbolic challenge involved in the project, because the previous efforts by the Quebec Bridge Company had ended in the collapse of the uncompleted bridge into the river, causing the death of 76 workmen. In order to reduce risks to a minimum, Dominion Bridge brought in the services of Canadian Bridge, from Walkerville, Ontario. Their alliance would result in the creation of the St. Lawrence Bridge Co. This new corporate entity built a plant in Rockfield, near the Lachine Canal, from which huge pieces of steel were carried to Québec by train. In 1916, both of the cantilever arms had been completed, and the 5,200 tons of the central span had been put on "six floating pontoons and then towed to the bridge, where they were to be lifted and affixed to the cantilevers using four hydraulic jacks." If this operation succeeded, it would constitute the greatest civil engineering feat ever accomplished in Canada. Unfortunately, a support component of the lift apparatus snapped, thus causing the main span to fall on September 11, 1916, plunging 13 men to their death. But only a year later, from September 17 to 20, a new span was successfully manoeuvred into place. After a series of tests, the Quebec Bridge was opened to rail traffic in August 1918. The following year, the St. Lawrence Bridge Co., which had since become the exclusive property of Dominion Bridge, changed into the Dominion Engineering and Machinery Company Limited and went into manufacturing industrial machinery.

* *cantilever*: a type of bridge whose main beams project beyond their support and which, in turn, hold up a beam of lesser span.

Chemicals and Carbochemical Products

Lachine Canal, Sluices on North Side of Raceway above St. Gabriel Locks, Looking North, National Archives of Canada, Department of Public Works collection, PA 110118, 1920

SKIN DEEP

For ordinary human beings, some products acquire more of a mystique than others. For example, the attraction of the locomotive or the ship is immeasurably stronger than any similar pull produced by paint or varnish. And yet, were it not for these protective coatings, a locomotive would quickly go to rust and a ship would rot out in no time at all. Conversely, developments in the paint sector have stemmed in large part from trends in transportation. Thus, the British perfected varnish in order to protect His Majesty's Ships. Likewise, without the orders of rail transportation companies, the United States industrial paint sector would have gotten off to a slower start than it actually did. In short, the transportation industry represented a clientele of choice for paint and varnish manufacturers. As it so happened, Montreal constituted a major pole for ship construction, the nation's largest centre for rolling stock manufacturing, and the hub of the entire Canadian transportation system. Given this situation, it was only natural that Canada's leading city, in terms of population and economic development, became the preferred location for the entire paint products sector, and that the Lachine Canal, with its strong concentration of transportation companies, constituted an irresistible magnet for the industry. Among the paint manufacturers established along the Lachine Canal, Canada Paint and its facilities on the St. Gabriel locks was the largest for a time. In 1895, the American multinational Sherwin-Williams settled several streets further to the west and subsequently bought out Canada Paint. In an exceedingly rare acquisition, the Canadian Walter Cunningham, the president of Sherwin-Williams, in 1911 bought up all British and Canadian shares of this multinational. Although this company did not remain under Canadian control after Cunningham's death in 1930, the paint industry in the canal sector nevertheless remained very much alive.

Lachine Canal, Little Weir above Côte St. Paul Locks from South Side of Canal, National Archives of Canada, Department of Public Works collection PA 109921, 1904

FROM GAS WORKS...

Unless the reader is particularly sharp-eyed, the buildings appearing in the upper left hand corner of this shot will escape unnoticed. But even for an expert, the outlines of the Consumer Gas gas holders are admittedly less than remarkable. Thus, in a way, this photo fails to do justice to the technological feat involved in extracting from coal a fuel source for producing light that was three times brighter and more intense than the flame of a candle. But while this photo may touch off few tangible memories, it nevertheless constitutes a reminder of a now-forgotten truth—that is, that manufactured gas and not natural gas reigned supreme in Montreal from 1837 to 1958. It takes considerable reserves of imagination to grasp the coal gas manufacturing process in all its complexity. With luck, however, a verbal description may help to make up for this difficulty. In brief, coal was heated to a high temperature in ovens and distillation *retorts** to produce coal gas. Upon exiting from the retorts, gas was drawn through a hydraulic main, or cylinder, surrounded by water that insulated the retorts from the rest of the circuit. Following this, the gas was directed through a cooler system (i.e., cooled tubing) in order to condense tar, and from there into a vat lined with lime and various other purifying materials. Finally, gas was collected in a gas holder, also known as a gasometer.[2] This method was also used during the 19th century by another large gas manufacturing plant, the Montreal Gas Company, located on Ottawa Street near the Lachine Canal.

* *retort*: an enclosed space made of refractory materials in which coal was coked—that is, distilled or decomposed.

2. DAUMAS, Maurice, *Histoire générale des techniques*, Vol. 3. *L'expansion du machinisme*, Paris: PUF, 1968, p. 638.

E. Bruce McLaren & Co., *View of the Battery Coke Side Bench Showing an Oven Being Pushed*, Montreal Coke and Mfg. Co., 1928

... TO COKING PLANTS

With the advent of electric lighting, gas lighting suddenly appeared less dazzling. After all, the flame of a gas lamp put out scarcely more light than a 25-watt bulb. Having both power and money in sufficient supply, gas light companies responded by tabling on the new technologies. They adopted the Welsbach mantle, which provided electric lighting without electricity, and subsequently discontinued the use of coal gas in favour of new types of manufactured gas. In Canada, carbureted water gas was chosen by several companies, including the LaSalle Gas Works in 1914. In brief, the process of producing it consisted first of loading *coke** into a cupola furnace and heating this material until it became incandescent through contact with air, so as to obtain a mixture of hydrogen and carbon monoxide. On account of its low heating power, this gas was *carbureted** in a carburetor, then sent on to a purifier, after which it was stored in gas holders. In Montreal, in 1925, LaSalle Gas, which was owned by the Montreal Light, Heat and Power Company, produced more than 127,426 cubic metres of gas using this method. Feeling increasingly confined within its quarters, the MLHP then decided to expand its gas plant and add a coking plant. To achieve this objective, it formed an alliance with the Koppers Company of Pittsburgh, which had designed and produced several coking facilities. It thus hoped to improve the quality of its water gas and obtain top-quality coke for use either in households or in blast furnaces and foundries. The Montreal Coke and Manufacturing Company, better known as LaSalle Coke was thus founded. In 1927, the company built 59 coking ovens capable of carbonizing 17 tons of coal in 15 hours, if domestic coke was desired, or in 24 hours in the case of metallurgical coke.

* *coke*: the solid residue produced by distilling coal at high temperature.

* *carburet*: to increase the available fuel energy of gas by carburizing it—that is, by combining or mixing gas with a gas hydrocarbon.

The Electrical Industry

Cotton Serving Department, Bell Canada Historical Collection, R. V. Holt / Northern Electric, Cat: 31008 s - 90 F: 28412, undated

AN ELECTRICAL INDUSTRY

During summer 1874, while at his father's home in Brantford, Ontario, *Alexander Graham Bell** produced a synthesis of his research on sound transmission. In July 1875, he succeeded in transmitting the first human sounds in his Boston workshop—although these sounds were practically inaudible. On February 14, 1876, Bell filed an application with the United States Patent Office for "Improvements in Telegraphy." On March 10, 1876, over a rudimentary telephone, Bell shouted to his assistant: "Mr. Watson, come here, I want to see you." On July 8, 1877, he founded Bell Telephone Company, and the following day transferred his Canadian rights to his father Alexander Melville Bell. In late 1879, the National Bell Telephone Company bought out Melville Bell's rights and founded the Bell Telephone Company for the purpose of marketing the invention. In order to avoid losing his patent, Bell began producing the magneto telephone in 1882. Within a decade, the Bell Company was using a paraffin-coated cotton sleeve to insulate the cables it manufactured. In 1895, Northern Electric was founded for the purpose of producing telephone components. Initially a luxury item, the telephone saw its market steadily expand. In 1902, fully cognizant of this trend, Bell built a large plant in Montreal, on Guy Street, in the vicinity of the Lachine Canal. The enthusiasm of Canadians for this means of communication grew unabated, with the result that, in 1913, Northern Electric built the continent's largest electric wire and cable factory on the south shore of the Lachine Canal, near the St. Gabriel locks. Beginning in 1916, the plant functioned at full capacity thanks to a workforce numbering in the thousands. Over the following decades, Northern Electric continually increased the production capacity of its main manufacturing centre and regularly enlarged its plant.

* *Alexander Graham Bell:* Alexander Graham Bell, born in Great Britain, had applied for American citizenship in 1874, but was naturalized only in November 1882. Legally speaking, Bell was thus a Canadian when he invented the telephone.

Interior View of Canadian Allis-Chalmers Limited Plant, Section of Main Machine Shop—Steam Shovel Built in 1906, National Archives of Canada, Andrew Merrilees collection, PA 171764, ca 1906. Although this photograph was taken in Allis-Chalmers' Toronto shop, the inside was similar to that of Lachine.

DAVID AND GOLIATH

At the turn of the 20th century, several American companies were busily carving out national empires through mergers and acquisitions. For these large industrial corporations, American markets suddenly appeared too small, and Canada, with its incomparable riches, seemed like a natural stepping stone on their path to world conquest. Obstacles to dumping aside, this country represented an easy prey, for although the Canadian government had erected a tariff barrier to protect companies, it prevented neither foreign takeovers nor the creation of branch firms. An unobstructed path stretched out before Americans since, in order to produce locally, they had only to buy up a Canadian rival or rely on their own resources. As a major industrial centre, the Lachine Canal was witness to both strategies, with some of its prize companies being swallowed up by American firms. Occasionally however, the formidable strength of American companies was not equal to the situation and, as in a rematch between David and Goliath, Canadian companies succeeded in bringing the giant down. Thus, when the major American manufacturer Allis-Chalmers-Bullock opened a plant in Lachine, all signs suggested that Canadian General Electric would be outmatched. Nevertheless, CGE, the owner of the Royal Electric's Montreal plant, proved to be a particularly daunting opponent. Admittedly, it had an established clientele and sold quality leading-edge electrical equipment thanks to its licences on American patents. In 1912, CGE took advantage of a reorganization of ACB's parent company to spirit away the Canadian rights on the American head firm's patents from the Canadian branch. The following year, the Lachine plant of Allis-Chalmers and its 550 employees fell under CGE's sway with scarcely a murmur of resistance. Under the name of Canadian Allis-Chalmers, the plant continued to produce electrical and non-electrical machinery.

Working in the Factory, Living in the City

L'Opinion publique, 2 November 1871

"My poor husband, you complain of your ten hours of work: I've been working for fourteen hours now and my day is not over yet."

Both the illustration and the accompanying caption published in the *Opinion publique* speak volumes about the late-19th-century working-class world. Obviously, the manifest issue is one of working hours, whether in a man's or a woman's occupation; after all, the horizons of women extended beyond the four walls of a dwelling. However, beyond the veiled allusion to the heavier burden falling to women, it is important to make out an entire web of working conditions.

The visual denotes other considerations specific to the working-class world. What first meets the eye is the cramped quarters and the absence of windows—hence, the absence of ventilation. Furthermore, judging by the meal on the table, this family's diet appears to be quite frugal. The baby, no doubt, denotes family and family size. And if there is one reality closely associated with working-class demography of the time, it is very much indeed the drop in the birth rate and the phenomenon of infant mortality. This mortality stemmed from a variety of causes, including unsanitary living quarters and the absence of fresh food and of fresh milk, in particular.

The question of mortality depended not only on diet but also on the physical working environment and housing conditions. From this perspective, the issue of the working-class habitat inevitably overlaps that of urbanization, which leads directly to the question of how urban living conditions either improved or deteriorated.

Work in the Factory: From Tolerating the Intolerable to the Quest for Quality of Life

In 1846, the industrial establishments along the Lachine Canal provided just a few hundred jobs, but some 25 years later 4,089 people were employed in the area. By itself, the area accounted for nearly one quarter of the artisans and workers active in Montreal's manufacturing sector—a proportion that would remain steady until the early days of World War II. At that time, Montreal was still the country's industrial metropolis, although Toronto was closing in behind it. The city was home to the largest worker population in the country. The hundred or so companies located along the canal had hired approximately 28,000 to work in their factories, while the island's other 2,000 companies struggled to create 90,000 manufacturing jobs. Thus, more than one employee in five on Montreal Island worked in one of the major industrial complexes making up the landscape around the Lachine Canal.

Whether workers numbered in the hundreds or thousands, all were confronted with the same reality. For, however many they may have been, they were all alike in having to sell their labour power in order to earn a living. In an environment in which the machine came first, the pay was not always enough to feed a man, much less his family. This predicament was all the more acute since the owners of the means of production continually strived to simplify the work of their employees until the latter ended up only executing well-defined, repetitive tasks, which by definition were less well paid. Ultimately, and above all after World War I, the elimination of wasted motion and the strict specialization of tasks would lead to the introduction of assembly line work in plants. Manufacturers wholeheartedly took up the mechanization of operations in their industrial facilities. They gauged this mechanical progress not only by the increasing complexity of the operations performed by machines, but also by the ever simpler running of the machines themselves.

Between 1846 and 1945, productivity skyrocketed, owing to a combination of rationalized production, an efficient division of duties, enhanced mechanization, and an accelerated rate of output. In that connection, the example of cotton manufactures is particularly eloquent. Thus, in 1850, it is estimated that a cotton spinner "watched over 4 looms of 36 inches, each operating at a rate of 112 beat-ups per minute; [and that] in 1925, the same worker had to monitor 20 looms or more, operating at a rate of 170 beat-ups per minute. Assuming that down time was equal from one technology to the next, the productivity of this worker increased 750 percent."[1] While this observation was drawn from the textile sector, could it accurately reflect manufacturing production as a whole? Without too great a risk of error, it may be asserted that productivity doubled, helping to increase the wealth of the nation.

As to who profited from this wealth, the answer seems a bit more complex. As consumers, the entire population gained, since the increase in productivity led to a drop in the prices of manufactured goods and, indirectly, those of food. As for workers or management, it is clear that managers reaped the main benefits, called profits, while, in comparison, workers received only a pittance, called pay or remuneration. As well, the distribution of wealth differed depending on one's position in the system. For those in management, wealth was calculated according to the number of shares they held in the company, whereas for workers it depended directly on their capacity to hold a job. During an era when the roles of each party were clearly defined, employees rarely received shares from their employers. In that respect, Montreal Rolling Mills stands out as a pioneer, since, in February 1906, it sold 350 shares to some 20 employees with considerable seniority for the sum of $125 per share. Current knowledge suggests that this is the first example of its kind to occur along the Lachine Canal, and most likely one of the first such initiatives, if not indeed the first, in Quebec and quite possibly Canada.

As labour was a commodity, a skilled, experienced worker enjoyed better bargaining power than did an unskilled worker. In a context founded on the law of supply and demand, a skilled worker was not, however, immune from dramatic drops in wages. The supply of labour had only to exceed demand in order for this same worker to be relegated to the low income category. Gender, age, community of origin, the size of the industrial facility, competition within a sector, and the degree of solidarity among workers were all criteria that influenced the price of labour in Montreal. Indubitably, another factor was rising unemployment, which created a reserve of manpower and thus contributed to lower wages.

Whether they were skilled or unskilled, many workers only just scraped by on their

1. Beat-up: One of the basic motions of weaving, in which weft yarn is forced into position in a fabric. McCULLOUGH, A. B., *L'industrie textile au Canada. Histoire et patrimoine*, Ottawa: Lieux historiques nationaux, Services des parcs, Environnement Canada, 1992, p. 20.

pay. The challenge was all the greater since their wages were far from being indexed to the cost of living. For example, in 1900, an average family of five needed $9.64 per week for its most basic needs, and $13.77 for all others. However, historian Terry Copp has estimated that during this era, the average income of men was $7.78 per week and that of women $3.65.[2] In this context, it is understandable that making ends meet was an overriding concern of employees, and that many women and children ended up working in the mills in order to make the next month's rent. Despite this contribution, working-class families lived under conditions of constant precariousness.

It was not uncommon for a company along the Lachine Canal to lay off its workers when circumstances forced it to do so, thus creating a situation of underemployment. In a society devoid of a genuine social safety net, this predicament very often engendered misery and distress among individuals and society alike. A less dramatic but more frequent tendency among canal companies was to cut back on their employees' working hours, or not pay a fair price for overtime. Likewise, the introduction of piecework in a factory not only spelled a meagre existence for its employees, but also rather quickly locked many members of a worker's family into an infernal cycle of living hand-to-mouth.

Under the threat of unemployment, many workers submitted to the arbitrary rule of management and a horrendous labour system. Nevertheless, many rose up against any loss of income. And their actions were all the more praiseworthy since, until 1872 at least, any striker was liable to fines and prison sentences for infringing on an employer's freedom of contract. Even afterwards, however, strikes remained an act of courage since, although the Canadian law of 1872 recognized the right of association, it in no way prevented management from calling in strike-breakers. During this period, in which management's arbitrary rule had the force of law and in which job security was a hollow notion, a great many strikers saw their hopes for a better life foiled by the hiring of scabs. For these men and women a strike necessarily entailed famine and poverty, especially when there was no strike fund. Occasionally, the solidarity of fellow workers managed to make an employer bend, particularly if the strikers were members of a union. However, despite the symbolic value of strikes, they were hardly a frequent event in the lives of workers in canal factories. In truth, labour conflicts were quite rare, occurring barely a hundred or so times among 600 manufacturing firms throughout 100 years of history.

Be that as it may, workers' conditions along the Lachine Canal cannot be reduced either to the wage relationship or the instability of family income. In effect, workers did not merely earn their living in a factory, they also spent most of their time there. Forced to put in long hours, men, women, and children often worked under unhealthful conditions and, as a result, occasionally contracted occupational diseases or suffered one of the many work-related accidents. Aware that they were killing themselves at the job, literally and figuratively, workers claimed the right to decent living conditions from managers. Their rights often disregarded, workers were sometimes pushed to declare a strike in order to obtain better working conditions. The Montreal Ten-Hour Movement of 1833, followed by the pan-Canadian Nine-Hour Movement of 1872, were turning points in the history of the country's labour relations, as they spearheaded the founding of the first labour unions and led to the decriminalization of strikes. Likewise, under pressure from the labour movement and public opinion, the government of Quebec was prompted to pass a series of legislative measures dealing with working conditions in factories.

2. Copp, Terry, *Classe ouvrière et pauvreté. Les conditions de vie des travailleurs montréalais 1897-1929*, Montréal: Les éditions du Boréal Express, p. 31 et 32.

While, at the outset, this working environment knew no rule other than the laws of market cycles and the fiats of management, it gradually became subject to a minimum of regulations enacted by the province of Quebec. Thus, factory managements deployed efforts to make workplaces safe and clean—albeit timidly at first. The provincial government would also make it difficult, but not impossible, for young children to work, and would strive to reduce the working hours of women and children. In 1909, it adopted a bill designed to shift the burden of proof from the employee to the employer in cases of work-related accidents. Ten years later, Quebec legislators adopted a minimum wage law for "factory girls," a measure that was not genuinely enforced until 1925. Considering the situation that had prevailed previously, it is certainly true that such social legislation ensured the working class a degree of well-being; nevertheless, in the context of the economic crisis of the 1930s, the legislation failed to address the social problems of the work-ready, such as their meagre wages, and disregarded those of the unemployable, who were often condemned to eke out a bare bones existence.

The Great Depression forced the federal and provincial governments into action. Beginning in 1930, Canada enacted legislation to assist the unemployed through the famous direct assistance measures and, in 1940, set up an unemployment insurance system. In Quebec, a law on decent wages was not adopted by the provincial legislature until 1937. All these initiatives contributed to the well-being of workers, and, in 1941, the working class was in a better position than it had been 100 years earlier—at least in terms of factory working conditions, for families found it as difficult as ever to make ends meet. In short, in 1941 as in 1846, workers led modest lives, but at least by the mid-20th century, the dark clouds of poverty had lifted.

City Living

The density of the factories along the Lachine Canal was such a prominent feature of the local landscape that, at least in the Montreal sector stretching from the mouth of the canal to Côte St. Paul, it is by no means a rhetorical flourish to speak of an industrial wall. As a structuring component of the urban landscape, these industrial plants were much coveted by municipal elected representatives, who were well aware of the role industrialization played in driving the economy and shaping society. These officials did everything in their power to attract industries to settle on their territory. And it is true that the siting of some of these factories locally triggered the urbanization of surrounding lands, as shown in the examples of Ville St. Pierre and Côte St.Paul. However, these new municipalities were founded only after the Church had raised new parishes within their boundaries. After being subjected to pressures from the clergy, they were forced to yield to the "dictatorship of industry" in some degree, since industrialization necessarily entailed the expansion of these factories and, hence, greater numbers of workers. These same workers, who continually demanded better working conditions, would also petition their municipal representatives for improvements to their environment.

At the time of their inception, these new urban environments were communities where the male component of the workforce was overrepresented—a characteristic of any new community then in the process of urbanizing. Twenty years later (that is, around 1881), the situation had stabilized, as women now made up 15% of the workforce in almost all of the municipalities along the corridor, with most of this number being concentrated in garment-making. French Canadians began arriving massively from the countryside toward the mid-1860s, and congregated along the north shore of the canal stretching from the St. Anne sector to the west end of St. Henri. They were also present in several enclaves along the south shore, making inroads into

Côte St. Paul, notably. The Catholic Irish settled in the St. Anne and St. Gabriel neighbourhoods, whereas Protestant Anglophones (Irish, English, or Scottish) confined themselves to the north end of St. Anne or the south end of St. Gabriel.

During the second half of the 19th century, these workers lived in the vicinity of their workplace, whether in Côte St. Paul or St. Henri. There was a strong correlation between the industries present in a given municipal territory and the trades practised by area residents, as has been confirmed by an analysis of occupations registered in censuses according to neighbourhood and block. Furthermore, this finding from various censuses also points to the stability of the day worker segment and a drop-off in labour from the lumber and leather trades—a decrease corroborated, moreover, by trends in the corporate names in evidence along the industrial corridor. As for the so-called mechanical trades—that is, of the kind encountered in the shops of the Grand Trunk Railway—these were filled massively by the Protestant Anglo-Saxons of St. Anne and St. Gabriel. Thus, there was a relationship between ethnic group, religion, trades, industries, and municipal territory.

These urban living conditions can often be summed up in a single word: insalubriousness. Miasmas and other unsanitary conditions in housing or factories produced a noticeably higher infant mortality rate and greater numbers of tuberculosis cases in working-class communities than elsewhere. In order to attract new taxpayers, some municipalities enacted measures designed to expel certain types of particularly nauseating or health-endangering industries, or at least push them away to remoter areas. However, attracting industries demanded a great deal from municipalities, which were forced to comply with industry dictates having to do with tax exemptions, various advantages, and improvements in urban transportation, to name but a few examples. In turn, these demands produced a serious impact on public finances, particularly in an environment in which tenants predominated.

Among the factors contributing to the overall betterment of urban living conditions were improvements to the wastewater system (or indeed the construction of a system), the creation of a number of city parks, and the introduction of electricity in many factories. Thus, between 1876 and 1938, tuberculosis cases dropped by three-quarters and the infant mortality rate fell considerably in the neighbourhoods bordering the canal. It is true that high mortality rates stemmed not only from a lack of proper hygiene but also from poor living conditions and, in particular, the spread of bacteria. On the other hand, the period was also witness to the emergence of solutions such as the "Goutte de lait," a network of milk stations that distributed fresh milk for babies—not to mention vaccination campaigns targeted at certain, specific diseases. When iceboxes began to be used in homes, food—and milk, in particular—could be kept fresh, at least for those who could afford one.

Whereas, in the 19th century, industrialization entailed the urbanization of the southwest portion of Montreal Island, to a point of even dictating its development, the picture partially evolved over the course of the 20th century. Workers continued to live in the vicinity of their workplace, but at least industrialists no longer imposed their fiats on elected municipal representatives. But to learn that lesson, municipalities had been forced to go into debt and be annexed, at least in the Montreal sector.

The World of Work

Working Conditions

DOMINION ILLUSTRATED, *Establishment of W. Clendinneng & Son, Interior View*, Bibliothèque nationale du Québec, E.-Z. Massicotte albums, 1891

IRONBOUND

Even though mechanization was not as relentless in iron foundries as in other sectors, moulders nevertheless saw their craftsmanship eventually supplanted by machine production. Ironically enough, machine moulders made out better than *bench moulders** and stove moulders. In Montreal, foundry workers combatted mechanization tooth and nail, while foundry owners did everything in their power to accentuate it. It was a lopsided contest. Language and religion divided moulders, and each strike lost diminished their capacity to defend themselves against the arbitrary rule of management. And although they had a union organization, it remained a fragile structure. Thus, in 1862, the Ives & Allen stove foundry single-handedly broke the Montreal chapter of the Ironmoulders Union of North America. Until March 1879, the Ives foundrymen had submitted to the company's conditions, but a 15 to 25% reduction in wages incited them to go on strike. By way of response, H. R. Ives simply hired strikebreakers. A combination of fights, police intervention, and "scabs" finally bested the workers. In an industry whose captains were known for their intransigence, William Clendinneng, the owner of a stoveworks, was something of an oddity. This man continually improved the working conditions of his employees, endowing his factory with a reading room (1872), inaugurating annual picnics (1873), and granting wage increases after only a single day of strike (1880). However, this progressive policy posed some risk for Clendinneng, as payroll expenses at his factory mounted while the same expenditures at the other foundries fell. In 1883, he too was forced to cut his men's wages. The strikes of 1883, 1886, 1889, and 1892 produced no improvements whatsoever. Clendinneng would not be budged, and the workers bitterly discovered that this friend of the workingman could be as iron-fisted an employer as the others.

* *Bench moulder*: during the mould production phase, a moulder reproduced small articles, such as machinery parts, on a moulding bench. Around 1860, a bench moulding machine was created for the purpose of reproducing mass quantities of products that had been previously moulded by hand.

Northern Electric Manufacturing Co., Blacksmith Shop, Bob (?) - Billy (?) in Foreground, Montreal, Bell Canada Historical Collection, Northern Electric Co. collection, 4465, undated

FROM INSALUBRIOUS INDUSTRIES...

On the Lachine Canal, home to an abundance of forge-owning companies and foundries, men and sometimes boys had to deal with not only the rudimentary hygiene of the 19th century, but also the intense heat, the smoke and noise, the metal dust, and the deleterious gases characterizing the metallurgical industry. To air out this unhealthy atmosphere, factory-owners used the most basic of ventilation methods (such as opening doors and windows, which would have been of dubious effectiveness). In this suffocating environment, workmen often had only fetid water with which to quench their thirst. And while they might somehow manage to restore their water intake, they did not always have enough time to allow their sweat-drenched work clothes to dry: employees often showed up for work—in freezing weather or dry—with their garments no less sopping than they had been the day before. As an editor of the *Commercial Sketch of Montreal* observed closehand during a tour of the Eagle Foundry in 1868, insalubriousness was a particularly striking feature of foundries. Like the *salamanders** of myth, these seemingly "incombustible" foundry workers filled moulds with molten iron, while shouts reverberated amidst explosions occurring every few seconds. "To add still more to the unnatural aspects of affairs, a dense foggy cloud of sulphurous steam enveloped the place." This editor, his mouth filled with the taste of sulphur and his lungs oppressed by the irrespirable atmosphere, beat a hasty retreat toward a source of fresh air.[3] As for the worker-moulders, no such opportunity was available to them, as they had to work hard—an average of 10 hours a day, six days a week—to earn a living. Although the installation of the first mechanical fans made life somewhat easier for moulders, the introduction of piecework and the infernal work pace that it imposed prompted one of these workmen, Richard Powers, to cry out in 1886 that his trade had condemned him to nothing less than a slow suicide.

* *Salamander*: a mythical creature once thought capable of living in or withstanding fire. Alternately, this name has been given to a variety of instruments associated with intense heat or fire, such as a poker, an overhead grill, a portable stove, and, in metallurgy particularly, a mass of solidified material, usually metallic, left inside a blast furnace. In French, "salamandre" denotes a slow combustion stove placed inside a hearth (although in Quebec, it was also called a "tortue" because of its turtle-paced fire). Interestingly enough, in English-speaking Canada, this same type of stove is occasionally referred to as a "Quebec heater."

Notice to readers. Although this photograph does not reflect the unsanitary conditions prevailing in the metallurgical sector, it does provide a glimpse of working habits in metal-working shops at the turn of the 20th century. Except for the photo of the Clendinneng operations, to our knowledge there is no other photo of the inside of a foundry located along the Lachine Canal.

3. *Commercial Sketch of Montreal and its Superiority as a Wholesale Market.* Montreal: Chisholm & Dodd, Publishers, 1868,pp. 27-29

Employees Pouring Metal, Operating Cupolas,
Canada Science and Technology Museum,
Canadian National collection, 41061-2, 1939

... TO SALUBRIOUS INDUSTRIES

In 1891, in a first for Quebec, the provincial government established a list of 28 types of manufacturing considered insalubrious and dangerous. In a "worst of" contest, the metal-working industry, represented by foundries, would have been runner-up, while the chemical industry would have taken "top" honours. The metallurgical sector owed this dubious distinction less to the dangers of shop floors than to the unhealthy air that clogged and intoxicated workers' bodies. And yet, all it would have taken to improve the situation was to sufficiently air workshops with centrifugal fans in order to expel deleterious gases, smoke, and metal dust, as well as to isolate unhealthy phases in the production process and install showers. In the case of foundries, which were heated with open fires (a major cause of noxious smoke and deleterious gases), heating systems also had to be installed. After releasing this list, the Quebec government, a champion of laissez-faire policies, allowed the situation to continue on unchanged until 1911, when it enacted legislation designed to eliminate the unhealthful conditions reigning in metal-working shops. In 1914, under the new regulations, Chief Factory Inspector Louis Guyon put a civil servant in charge of inspecting foundries, munitions factories, rolling mills, and other iron and steel industries. This new inspector advocated a series of measures with which most companies complied, all the more so because the war economy provided them an incentive to invest in upgrading their facilities. Thus, by 1919, almost all metal-working shops had become equipped with showers, fans, modern heating systems, lockers, and dining areas closed off from work stations. Metallurgy, once an insalubrious industry in which workers traded away their health in return for wages, became a healthier sector in which workers were less likely to endanger their quality of life. Such progress remained tentative, however, because many savings-minded employers used their ventilation devices sparingly—which explains why compliance had to be continually monitored.

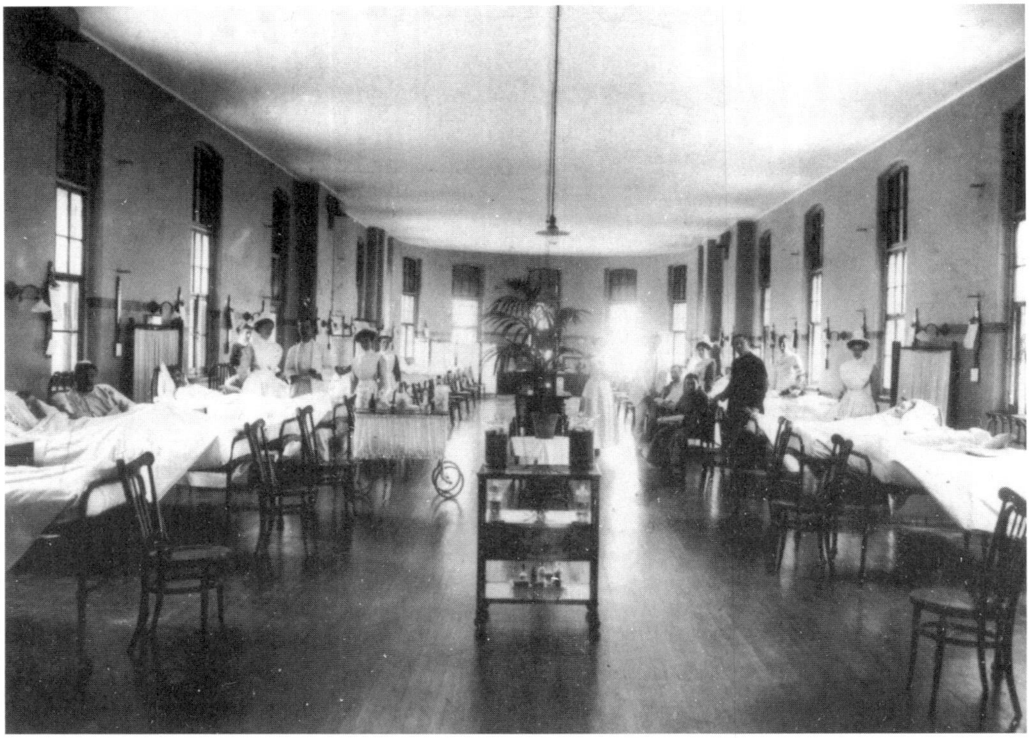

Patients, Ward, Montreal General Hospital,
McGill Archives, PR 000512, undated

BREATHING EASIER

The violence of an industrial accident makes it a spectacular event, stemming from identifiable circumstantial causes. In comparison, an occupational disease is a process in which the victim's health declines dramatically on occasion, but most often deteriorates insidiously through contact with a toxic agent, a virus, or an unhealthy atmosphere. Thus, a worker's health could be destroyed—and indeed his or her very life imperilled—by handling toxic agents such as lead, mercury, arsenic, and phosphorus; by contracting a virus, such as anthrax; or by inhaling deleterious gases, hydrocarbons, and unwholesome dust. In Quebec, measures to prevent or eliminate occupational diseases were slow in coming. Obviously, health officers and factory inspectors strived to improve the sanitary conditions of factories, in particular by requiring restrooms, instilling notions of public health among workers, and promoting improved ventilation as a means to purify the filthy atmospheres that filled the industrial workplace. In reality, they could deal only with the most urgent of the problems at hand. Government intervention dates to 1890, when it prohibited boys under 16 and girls under 18 from working in those industries that had been deemed insalubrious or dangerous. In theory, the young working population would no longer be exposed to the fetid odours emanating from slaughterhouses, the noxious gases released by acid factories and white lead manufactures, the metal fumes discharged by foundries, or the harmful dust and vapours of chemical suppliers. In 1902, the government expanded its list, and from that time under-age youths were forbidden from working in 64 industries. That same year, the American labour movement pushed to make occupational diseases subject to mandatory reporting; the "Inspection of Factories" unit took up a similar demand in Quebec some time later.

Lachine Canal, View of Canal, Looking East, Taken from Top of Charlevoix Street Bridge (Brewster's), National Archives of Canada, Department of Public Works collection, PA 119398, 1920

THE SPECTRE OF POVERTY

As there was no public insurance system in Quebec prior to 1909, and scarcely any private system to speak of, a work-related accident, if serious, was a sure ticket to poverty. Of course, the Lower Canada *Civil Code* of 1866 empowered workers and their inheritors to take employers to court. But prior even to proving to a judge that damage had been suffered at the hands of an employer, an accident victim first had to find an attorney (called an advocate in Quebec). And given their own fee requirements, advocates only accepted cases they had a chance of winning. Thus, few of them took on cases in which the accident victim appeared to have lacked foresight or been negligent. Likewise, they were little inclined to counsel a worker to bring proceedings for damages if there was no witness to the accident, since "anonymous" cases had the reputation of being lost causes. An experienced worker would probably be advised to go no further, because in such cases, the judge no longer assessed the employer's share of responsibility in an accident, but instead determined whether the employee had worked according to the established rules of the trade. For the select few who managed to persuade an advocate to take their case, initiating an action entailed embarking on a legal merry-go-round as, more often than not, they would be dragged from one courtroom to another by representatives for management who exhausted all procedural avenues in an effort to discourage the complaining party. When confronted with such delaying tactics, many an unemployable accident victim ended up accepting an out-of-court settlement. Against this backdrop, it is indeed astounding that prior to 1911, 125 accident victims managed to obtain a judgment establishing the civil liability of their employer. Even so, not all of the injured were awarded damages since, according to the statistics established by Katherine Lippel, the Quebec Superior Court ruled in favour of employees in only "72.3% of cases, the Queen's Bench in a proportion of 78.6%, and the Supreme Court of Canada at a rate of 53%."[4]

4 LIPPEL, Katherine, *Le droit des accidentés du travail à une indemnité: analyse historique et critique*, Master of Law thesis, Université de Montréal Faculty of Law, 1986, p. 63.

Grand Trunk Railway, Point St. Charles Shops, Montreal,
National Archives of Canada, Andrew Merrilees
collection, PA 202580, 1880

AN OUNCE OF PREVENTION

Through continual contact with machines, workers were frequently exposed to danger. However, economic interests inclined employers to increase the speed of machines without also investing in preventive measures. The result was an upsurge in work-related accidents. With public opinion in an uproar over the misery experienced by workers, the Canadian government established a board of inquiry on factories in 1882. During the hearings, the commissioners were taken aback to hear that no guards whatsoever were placed around gears, lever arms, pulleys, belts, and steam machines, and that several girls had had their hair caught in drive shafts and been killed as a result. But above all, the commissioners severely blamed factories for failing to protect workers adequately. In 1885, Quebec passed the *Act to protect the life and health of persons employed in factories*, thus implementing a labour code. This law contained measures applying specifically to the manufacturing sector and provided the legal framework authorizing the Inspection of Factories unit, established in 1889, to act in the field of occupational health and safety. It was about time such measures were enacted, since the number of accidents had been rising steadily and a number of workers had been killed or mutilated after their clothing became caught in a belt or a drive shaft. During the 1890s, inspectors strove to make industrial machinery a great deal safer, an objective that was emphatically supported by insurance companies. As early as 1892, in an indication of the improvements to come, the Grand Trunk Railway shops in Point St. Charles earned praise from provincial inspectors for their meritorious efforts toward achieving workplace safety. And, for companies requiring more stick and less carrot, in 1893, the Supreme Court of Canada set out the obligation of employers to take all precautions necessary to preventing accidents. This decision spurred many to action—to such an extent, in fact, that by 1902, the absence of protective gear on machinery was no longer the primary cause of work-related accidents, at least in the major urban centres.

Inside View of the Dominion Bridge Co., Limited,
National Archives of Canada, Dominion Bridge
collection, PA 102720, undated

INSURED VERSUS UNINSURED

In order to protect the industrial accident victim from a life of total poverty, the Grand Trunk shops eventually (some time prior to 1889) required their employees to form up a mutual assistance society. Under this policy, workers were indemnified in the event of illness or work-related accidents. Other companies, such as the Allan Lines prior to 1889 and Dominion Bridge most likely in 1896, began withholding a portion of their workers' salaries in order to take out life and accident insurance for them. Despite such initiatives, in 1889, during the hearings of the Commission on the Relations of Capital and Labor in Canada, workers from the Grand Trunk Railway and the Allan Lines spoke out against the control these companies exercised over the accumulated funds and denounced the difficulty they had encountered trying to obtain indemnities. The commission's reaction to this situation was to recommend that the Canadian government pass workers' insurance legislation, but to no avail. At any rate, neither workers nor employers wanted an insurance system that would require them to assume the better portion of costs—in the case of workers, because contributions would reduce their means of subsistence, and in the case of employers, because contributions would increase their production costs and reduce their profits accordingly. In the meantime, the number of industrial accidents continued to climb. As a result, labour organizations clamoured for a bill establishing the responsibility of employers, while manufacturers viewed lawsuits as being expensive and vexatious and for that reason as constituting a major irritant in the conduct of their business. In order to preserve labour peace, in 1907, the Government of Quebec set up the Globensky Commission to study the apportioning of responsibility in the event of industrial accidents. In the conclusions of its proceedings, the commission recommended paying out indemnities to accident victims and having this compensation covered by the insurance policy of employers. In 1909, the Government of Quebec adopted "a law designed to ensure that compensation was paid injured workers without regard to the identity of the responsible party," but did not, however, institute a compulsory insurance system.

Working Hours

Grand Trunk Railway, Old 4-4-0 Express Locomotive / Grand Trunk Railway in Erecting Shops, Point St. Charles, National Archives of Canada, Andrew Merrilees collection, PA 202438, 1859

THE NINE-HOUR MOVEMENT

Workers, like farmers, laboured from dawn to dusk. In addition, they performed their work according to the pace set by machines in environments that were often dirty and noisy, under the close supervision of foremen. Workers had to show up for work very early in the morning or risk being fined; typically, they rose around 5:00 a.m., breakfasted quickly, and then rushed off to the factory—in winter as in summer. For the next six to seven hours, they remained at their workstation without interruption until breaking at noon for dinner. During the short half hour allotted them, they ate their meal on location, and then returned to their duties, sometimes until 6:00 p.m. and even nightfall. This was the system that workers endured Monday to Saturday. Paid wages that barely ensured their survival, Montreal workers hesitated to mobilize against this situation, all the more so because any striker could be accused of illegally conspiring against the freedom of trade. Nevertheless, unrest was rampant, and in 1872 thousands of Montreal workers, including many from the Grand Trunk shops and the bronze foundries, joined the Nine-Hour Movement, which had been started in Ontario, with the objective of obtaining a shorter working day. In June, the Montreal chapter served employers with an ultimatum: give in on this demand or face a strike. Foundry worker-moulders, who were deeply involved in working-class protest, paralyzed five foundries. Thirty-five days later, the workers at E. Chanteloup, R. Mitchell & Co., and C. Garth & Co. won their demands, while those in another company succeeded in obtaining a pay increase. Despite these victories, the Movement collapsed in Montreal in August, and in Ontario a short time later, but not without first having bequeathed to the Canadian working class a law that "shielded labour unions from prosecution for illegal association with the intent of limiting the freedom of trade."[5]

5. Rouillard, Jacques, *Les travailleurs du coton au Québec, 1900-1915*, Montreal: Les Presses de l'Université du Québec, 1974, p. 28.

On the Lachine Canal, Anglin-Norcross, Ltd., McCord Museum, Notman Photographic Archives, MP 2083 (4), 1918

TEN, NINE, EIGHT...

Forced to put in backbreaking hours on the job, workers became exhausted and often fell sick. In the spring of 1833, Montreal carpenters had had enough and decided at a meeting that "ten hours of work a day is all a man is able to do." As a strike loomed closer, the master carpenters gave way, only to attempt to recover lost ground the following year by adding another hour to the working day. This backtracking prompted the carpenters to walk out. The movement spread to workmen in the other building trades, such as the stonemasons, and perhaps even the foundry workers-moulders. Although their strike failed, their demand for shorter working hours did not fall on deaf ears in a country where people often worked 12 hours or more per day, six days a week. Nevertheless, it was only in 1872 that this demand again led to a strike in Montreal. As is often the case, victory was short-lived, because in 1875 the foundrymen at the Mitchell and the Chanteloup works again agreed to work 10 hours a day. In 1889, the Commission on the Relations of Capital and Labor in Canada stated that workmen worked an average of 10 hours a day in Quebec factories. Actually, a great many workers, including employees in the cotton mills, began working at 6:30 a.m. and, after the midday break, continued at their tasks until 7:00 p.m. and sometimes 9:00 p.m. Yielding to this infernal system, workers generally suffered in silence. Sometimes, however, they rebelled. In October 1891, the 495 employees of Merchants Manufacturing, including 318 women, demanded a 60-hour working week, but were defeated. In June 1894, the carpenters, who were members of an international union, returned to centre stage, but their strike was no more successful than the one organized at Merchants. However, toward the end of the 1890s, the cigar makers' union opened a new chapter in Montreal working-class history when it obtained an eight-hour day in most cigar factories.

Montreal Dry Docks and Ship Repairing Company,
Parks Canada, Quebec Canals, 1919

ONE STEP FORWARD, TWO STEPS BACK

To the cry of "Don't cut men or wages, cut hours," the powerful Montreal local of the cigar makers' union won an eight-hour day in the 1890s, and then lost it during the strike of April 1900. Over the next several years, the labour movement adapted its demand according to the working conditions in force in a given factory or a given trade. This was because it appeared difficult to persuade an employer of the reasonableness of an eight-hour day when the majority of workers still put in a full 10 hours. This stratagem explains why, out of 24 strikes reported in the *Labour Gazette* between 1900 and 1914 for which a decrease in working hours was a major sticking point, workers struck as often for a nine- or 10-hour day as for an eight-hour day. But whether it was because of strikes or other factors, the number of firms where employees worked a 60-hour week declined from one year to the next. When World War I broke out, however, workers experienced a second setback. As soldiers of industry, they were not in a position to demand working fewer hours when so many men in uniform were sacrificing themselves for the nation. As a result, it was not uncommon to see workers breaking their backs at their machines for up to 13 to 14 hours a day. But as victory drew nearer, workers increasingly balked at maintaining this labour system. In May 1918, the 920 workmen of Fraser, Brace and Company and 350 of their fellow workers at the Montreal Dry Docks and Ship Repairing Company had had enough. They called a strike and abandoned work on the wood or iron ships ordered by the Canadian government from these Lachine Canal shipbuilders. By the end of their strike, the Montreal Marine Trades Federation had obtained a nine-hour workday for its membership. After this, the labour movement advocated a 44-hour workweek consisting of five days of eight hours, plus four hours Saturday morning.

Working Women

Colonial Bleaching and Printing Company,
National Archives of Canada, Dominion
Textile collection, PA 211417, ca 1906

TAILOR-MADE GHETTOS?

In 1871, the 380 factories on Montreal Island employed more than 20,000 people, one third of whom were women. However, one should not be misled into thinking that the entire female labour force worked in factories—particularly in the case of the garment industry where it was a widespread practice for milliners, seamstresses, and garment workers to labour away at home for a pittance. Excluding the needle trade, close to 4,000 women were employed by one or another of Montreal Island's industrial firms. This workforce was concentrated primarily in the shoe and boot factories, tobacco factories, and rubber goods plants, with 40 other industries, including tanneries and cotton mills, sharing the remainder. Curiously enough, neither the Canada Sugar refinery (Redpath) nor St. Lawrence Sugar hired women to any great extent. A decade later, 11,000 women and 20,000 men worked for manufacturers or storeowners. This upsurge in women's employment owed primarily to the development of the *sweating system** current in the needle trade and to the growth of shoe companies, which alone hired half of the female workforce. As for other sectors, the female labour force was concentrated primarily in cotton mills, tobacco factories, and silk and woollen mills, whereas male labour was omnipresent in metallurgy, transportation equipment, flour mills, sugar refineries, and sawmills. In 1891, 1,604 factories provided jobs to 30,000 men and 10,000 women. The increase in men's jobs stems in large part from the performance of the iron/steel and transportation equipment sectors. The decrease in women's employment can be put down to reduced hiring in the shoe industry and, above all, the replacement of seamstresses by male sewers in the garment industry. In fact, had it not been for job growth in cotton mills, particularly following the arrival of the Merchants Manufacturing Co., and steady employment levels in the tobacco products sector, the number of female workers would have dropped off even more precipitously.

* *Sweating system*: otherwise known as the piecework system, the sweating system worked as follows: "some operator or another would take out a contract in a factory—for producing 200 pairs of pants, for example. He would then divide this work up between 12 or 15 female workers, who worked for an outrageously low price. The women performed this work either in their own homes or, as was more often the case, at the contractor's premises. Working an average of 75 to 80 hours a week, the female employees, teenage girls for the most part, earned 50 cents to three dollars per week in 1895, rarely more."

6. Québec, *Rapport général du commissaire des Travaux publics de la rovince de Québec*, Quebec: Imprimé par Charles-François Langlois, 1895, p. 56.

Side View of Spinning Frames, National Archives of Canada, Dominion Textile Co., Ltd. Annual Report for Year Ended March 31st, 1928 / Part of File "H.S-4" Annual Reports, Pt. 1 1905-1949

MANPOWER—A NOTION THAT SPEAKS FOR... HIMSELF!

As during the previous century, women continued to be concentrated in the needle trade, the textile industry, and the leather and tobacco sectors, although they had made new inroads into such sectors as typography, chemical products, and electrical appliances. The waves of mass layoffs during the 1930s completely overturned the employment profile, and women were the first victims of this development. For one, employers began cutting back their workforce by dismissing wage-earners who were not the main family breadwinners. At the time, women's wages were primarily a form of family income supplement. Second, industries having a strong female labour force, such as the cotton mills, occasionally used the Great Depression as a pretext for replacing a woman paid the minimum wage by boys who were hired for ridiculously low wages—a complete distortion of the intention underlying minimum wage legislation. In reaction to this ploy, in 1932 the Women's Minimum Wage Commission petitioned Quebec to take action. But the provincial government only formally outlawed this practice in 1934, as the first signs of recovery began to appear on the horizon. There are many other examples of loopholes in this legislation, but the main point remains that women's wages always mirrored the vicissitudes of the market. During periods of growth, employers were willing to pay a decent wage, but in times of economic crisis, the old habits quickly regained the upper hand. As evidence of this observation, in 1934, women again began to be hired, albeit timidly so, and their base weekly wage began to climb. In short, the employment market was improving. Five years later, the nation went to war and industries called up their troops for duty. Whether married or single, Montreal women answered the call in great numbers; without them, the industrial front would have collapsed. When times of peace returned, they were urged to return to their kitchens, but nothing was able to erase this new awareness of their importance. That, however, is another story.

Weave Room, General View of Automatic Looms, National Archives of Canada, Dominion Textile Co., Ltd. Annual Report for Year Ended March 31st, 1928 / Part of File "H.S-4" Annual Reports, Pt. 1, 1905-1949

"FACTORY GIRLS"

During the between-war period, the typical Montreal working woman was young and unmarried. From Monday to Saturday, she worked long hours, continually attempting to keep up with the pace of the machines in cotton mills or garment factories. And, as she was rarely offered promotions, her horizon often extended no further than the same machine, year after year. As yet unmarried, she stayed with her parents. Paid on a piecework basis or an hourly wage, her earnings amounted to little. She often gave her family her meagre takings in order to put food on the table. If ever she wanted to leave her parents' home, she would be immediately confronted with the major problem of finding housing in line with her limited financial resources. Obviously, she could stay in a boarding house, but to do so she would have to cut back on such essentials as food and clothing. In order to leave home or escape poverty, another avenue would be to marry a man whose wages were often her own. While "some married women were forced by poverty or illness to take a job,"[7] most often marriage signified expulsion from the employment market. In the garment, tobacco, textile, shoe, paper bag, or electrical apparatus and supplies industries, this reserve of cheap labour represented quite a bargain for employers. However, it would be a mistake to think that women accepted such a situation impassively. For it so happened that their patience could be frayed to the point of snapping—so often, in fact, that between 1919 and 1939 women were at the centre of close to one half of the labour disputes that took place in Montreal. In terms of actual numbers, the revolt of the "*midinettes*," or female dressmakers, was the runaway winner with 125 of 320 strikes, followed by the shoe workers with 13, women textile workers with seven (including the strike of thousands of Dominion Textile employees in 1938), and female tobacco factory workers with five.

Intermediate Turn Stage in the Manufacture of a 75-mm. U.S. Shell, Dominion Bridge Co., Ltd., National Archives of Canada, Department of Defence collection, PA 186264, ca 1916-1917

WOMEN AND CANADA'S WAR PRODUCTION

When Canada entered the war in August 1914, it had not only to form an army, it also had to supply its troops with munitions, food, clothing, and materiel of all kinds. The scale and diversity of the Lachine Canal factories meant they would quickly assume a leading role in the industrial front then opening. During the next few years, canal manufacturers mobilized all of their resources, increased their production capacity, and built new factories. To accomplish this, they called back thousands of workers who had been laid off in the summer of 1914 and hired thousands of new employees, including a great many women. On that score, munitions manufacturers such as Montreal Ammunition stood out on account of their inclination to hire women. The integration of hundreds of women into the workforces of Dominion Bridge, Canadian Car and Foundry, the Grand Trunk shops and several other steel-making and rolling stock factories was just as remarkable. For in these industrial sectors from which female wage-earners were traditionally absent, women began to hold jobs—particularly after the 1917 conscription—that had until then been considered too physically or technically demanding for them. Thus, for example, Dominion Bridge unhesitatingly turned to women for precision work, and even paid them according to the principle of equal pay for equal work. On that point, however, this builder of metal structures was in a league by itself, for as a rule, the other Lachine Canal manufacturers paid women lower salaries than they did men. With the return to peace in 1918, the majority of female workers—and many men as well—were given their pink slips.

Women Operating Cartridge Case Presses, Canadian Allis-Chalmers, National Archives of Canada, Department of Defence collection, C 018864, ca 1915

Child Labour

Employees of the Merchants (Dominion Textile),
St. Henri Historic Society, 42-ph-2, 1905

CHILD LABOUR

While child labour is certainly not a recent phenomenon, it did in fact expand with the advent of the manufacturing industry. On the one hand, new, easy-to-operate machines entered the workplace, thus making it possible to supplant adults with children in the production of countless goods. On the other hand, many industrial establishments deliberately hired children in order to drive workers' wages down. In 1871, children represented a prime pool of manpower for Montreal cotton mills, printing firms, and tobacco factories, with girls being the hirees of choice for the needle trade, and boys the preferred candidates for jobs in foundries, steam engine shops, nail works, and rolling mills. Ten years later, the number of children working for manufacturing firms had risen from 3,213 to 3,626. While metallurgy, clothing, textile, tobacco and shoemaking accounted for a large proportion of such hiring, the other industries did not trail far behind. In 1882, a federal commission shed light on the difficult conditions to which children were subjected in factories. The long workdays, filthy environment, work-related accidents, absence of restrooms, noise, and pollution were all factors that alarmed public opinion. In 1885, the province of Quebec banned female and child labour beyond 12 hours a day, or 72 hours per week, and outlawed the hiring of boys under 12 and girls under 14. Despite this and subsequent legislation, the situation for children remained sordid. In 1911, a former spinner asserted that child labourers endured a plight worse than that of prisoners. For if by chance they managed to flee this "modern-day version of a prison whose steel bars were shut fast by the employer's greed and rapaciousness, their escape would result in extreme poverty and hunger for a great many of them."[8]

8. Rouillard, Jacques, Les travailleurs du coton au Québec 1900-1915, Montreal: Les Presses de l'Université du Québec, 1974, p. 59

WILSON, Stanley B., *Chicago Sewer Excavator*, National Archives of Canada, Dominion Bridge collection, PA 210926, 1906

THE MINIMUM LEGAL WORKING AGE: REALITY VERSUS FICTION

In theory, beginning in 1885, no boy under 12 and no girl under 14 had the right to work in a factory. In practice, however, the law did not have sufficient teeth to put an end to hiring even younger children. A child had only to provide a birth certificate in order for an employer to refrain from checking his or her background any further. While inspectors occasionally succeeded in expelling dubious cases, they were often unable to dismiss children who had successfully disguised their age using doctored certificates. This tactic was encountered primarily among girls, as the lower legal working age for boys left them with little incentive for attempting to break the law. But over and beyond these infractions, the inspectors opposed hiring at such an early age since it prevented the young working population from receiving a basic education. Inspectors' reports were fairly littered with appeals to public opinion; for example, in 1892 James Mitchell said that he had seen children "wallowing in ignorance, without having received so much as a smattering of instruction."[9] Among the various measures considered, the commissioners advocated raising the minimum legal age for working in a manufacturing firm and sending illiterate children back home. In 1902, Quebec raised the minimum legal working age of boys to 13. In 1907, it set 14 as the minimum age for boys and girls, and stipulated that illiterate children under 16 had to take evening courses if they wanted to keep their job. This measure represented a hollow commitment, since it was practically impossible to effectively monitor the attendance of children in evening schools. In 1910, a new bill authorized the inspector to administer a test to all working children under the age of 16 and to have them dismissed if they failed. Despite this law and its subsequent amendments, inspectors continued to be confronted with the widespread use of "convenience documents" and the difficulty of enforcing the legal provisions applying to evening school attendance.

9. Québec, *Inspection des manufactures. Rapport de James Mitchell. Rapport du commissaire de l'agriculture et de a colonisation de la province de Québec*, 1892, Québec: Imprimé par Charles-François Langlois, 1891, pp. 123-124.

Folding Department, Colonial Bleaching and Printing Company, Late 1800s, National Archives of Canada, Dominion Textile collection, PA 211411, ca 1906

AN INCOME SUPPLEMENT

In the working-class world, work often went hand in hand with misery and poverty: seasonal, cyclical unemployment was the rule; workplace accidents proliferated; filth and unhygienic surroundings reigned both in the city and in factories; and piecework failed to provide enough money to live on. Against this backdrop of hardship, it was by no means unusual for a family, particularly a day labourer's family, to only scrape by. In a world devoid of any social safety net and in which charity alone was able to offer temporary relief, parents, particularly those having few marketable skills, were sometimes unable to provide their families with subsistence. Many children had little choice but to seek employment. Some worked without leaving the family household, sewing garments with their mothers for manufacturers, while others left home early in the morning to slave over a machine in a cotton mill or a shoe factory, or near a foundry blast furnace. Although the actual worksites varied, these children had in common the characteristic of working hard for little—and above all, for less than what their parents earned. During an era when the low wages paid children were often rationalized as being appropriate because they were trainees, this abundant reserve of manpower certainly contributed to keeping the wages of day labourers and unskilled workers, female or male, at barebones levels. A vicious circle became entrenched, since "children [...] were put to work on even the hardest tasks, in damp and unhealthy areas and surrounded by dangerous machines [and], assuming they had managed to avoid mutilation, in the absence of education and despite great natural talents, they would never hold more than a subordinate's position."[10] In the role of day labourers, these children would, upon achieving adulthood, be confronted with the same difficulty of providing for the needs of their families and would, in turn, send their own children to work in the factory in order to improve the family finances.

10. QUÉBEC, *Inspection des manufactures. Rapport de Louis Guyon. Rapport du commissaire de l'agriculture et de la colonisation de la province de Québec*, 1891, Québec: Imprimé par Charles-François Langlois, 1891, pp. 123-124.

Urbanization of the Corridor

Working-class Communities

W. M. Notman & Sons, *Montreal Views from Street Railway Power House Chimney*, McCord Museum, Notman Photographic Archives, MP 2936, 1896

FROM MUNICIPALIZED OUTSKIRTS TO INDUSTRIALIZED SUBURB

The establishment of industries along the edges of the Lachine Canal had a significant impact on the urbanization of southwest Montreal Island. In the strip extending from the districts that at one time lay just outside the city fortifications to Côte St. Paul, residents settled close to their place of employment on both sides of the canal. During an era when urban mass transportation did not yet exist, clustering was the only way to answer the "siren song" of factories. Between 1861 and 1881, the population along this segment of the canal doubled. The raising of new parishes, following bitter internal wrangling among the ranks of the Catholic clergy, led to the founding of new municipalities such as St. Cunégonde and St. Henri on the north shore and St. Gabriel and Côte St. Paul on the south shore. In 1881, the municipalities of St. Cunégonde, St. Henri, and St. Gabriel had a population of 15,000, including more than 6,000 in St. Henri. Côte St. Paul was somewhat different; being a company town, it urbanized quite slowly. Thirty years later, the territory of St. Henri was home to more than 26,000, the great majority of whom were French-speaking and Catholic. With its population of 11,000, St. Cunégonde achieved a degree of urban development surpassing that of the other, peripheral communities of that time. Between 1887 and 1910, all four municipalities were swallowed up by the City of Montreal which, as Mayor Préfontaine noted, had an annexationist strategy "in order to effectively compete with the major American cities." In the meantime, Verdun, an enclave surrounded by the St. Lawrence River and the municipality of St. Gabriel, became a bedroom community, since it had no manufacturing industry within its immediate limits. It grew by leaps and bounds, eventually becoming the second most populous city on Montreal Island; between 1901 and 1931, its population rose from 1,900 to 61,000, although this development was to a great extent checked by the Great Depression.

Dominion Plant, Canada Car & Foundry Co., Ltd.,
National Archives of Canada, Andrew Merrilees collection, PA 202435, undated

WEST OF CÔTE ST. PAUL

From 1846 to 1940, the urbanization process spread throughout the low-population territories located in the western portion of the corridor, as these sectors became increasingly industrialized. Ville Saint-Pierre saw its population increase fourfold, rising from some 500 residents in 1901 to 2,200 in 1911. It benefited from the establishment of Canadian Car and Foundry and Consumer Glass during this same period. The village of Lachine officially became a municipality in 1848, acquired the status of town in 1872, and was made a city in 1909. Likewise, its population rose from 1,696 residents in 1871 to 10,699 in 1911 and 20,051 in 1941. Beginning in the 1880s, Lachine became a major industrial pole thanks to the siting of Dominion Bridge within its limits. Other industrial actors, such as Dominion Wire and Jenkins Valves, further added to the development of this forerunner to the contemporary industrial park. The third-ranking municipality in this sector, LaSalle, was a long time developing, as its population only progressed from several hundred residents in the early 20th century to 4,651 at the outbreak of World War II. These cities perpetuated the urbanizing trend that had previously occurred around the Lachine Canal and which was characterized by small, heavily industrialized communities and a large working-class population. Between 1871 and 1911, the municipalities along the canal corridor altogether accounted for 20% of the total population of Montreal Island, and were the place of employment for at least as many of the metropolis's working people.

Valentine & Sons' Publishing Co. Ltd., *St. Gabriel and St. Charles Churches, Point St. Charles*, Bibliothèque nationale du Québec, BNQ collection, CP 2726, undated

THE CANAL, A CULTURAL DIVIDE?

During the 19th century, Montreal received most of the immigrants settling in Lower Canada. Two thirds of these newcomers lived in the west of the city, with the result that, in 1831, Montreal had become a city with a majority English-speaking population—a phenomenon that intensified with the arrival of great numbers of Irish during the next two decades. The constant influx of French-speaking migrants from the countryside beginning in 1865 resulted in Montreal Francophones becoming a majority once again in 1871. However, in 1911, 11% of the Montreal population was neither French-speaking nor English-speaking: Montreal had begun to acquire its cosmopolitan character. The population of the southwest sector of the island was a part of this trend. In a sign of the times, Italian immigrants settled within the limits of St. Cunégonde and Lachine between 1891 and 1906. In the late 19th century, the St. Anne ward located at the Montreal mouth of the canal had a mixed population base, split in thirds between Irish, French Canadians, and English. On account of the high proportions of Francophones inhabiting the municipalities of St. Cunégonde and St. Henri, a French-speaking strip formed along the north side of the canal. Sizeable numbers of Irish settled in Griffintown, in the outlying area of the canal along the north side. The expansion and industrialization of the sector lying south of the canal coincided with the arrival of Canada Sugar and the Grand Trunk Railway shops and spurred new settlement in this area. Thus, these industries gave birth to the village of St. Gabriel, whose northern portion bordering the canal was inhabited by Catholic Irish, whereas the south attracted—not without some friction—English, Irish, or Scottish Protestants. The French Canadians formed a kind of buffer group sandwiched between these opposing ends. During the 20th century, the city of Verdun, an outgrowth of St. Gabriel, was home primarily to Protestant English-speakers up until World War II, as the result of a heavy influx of British immigrants during the 1920s.

Arless, Sr., Richard, *Point St. Charles District, Montreal, P. Q.*, National Archives of Canada, Richard Arless collection, PA 043877, 1946

MUNICIPALIZATION AND CITY DEBT

The marked urban density of southwest Montreal Island resulted in part from the decisions and actions of municipal leaders who were directly involved in developing the industrial use of their respective territories. A dual trend was set in motion with first the granting of tax credits and abatements in order to attract industries and then the building of urban infrastructures—a trend that affected St. Henri as much as it did St. Cunégonde and even St. Gabriel, not to mention Lachine and Ville Saint-Pierre. St. Henri, following St. Gabriel's example, exempted manufacturing firms from municipal taxes for 10 years. Thanks to this measure, it was able to attract Merchants Manufacturing, a cotton goods manufacturer that turned economic rivalry to good account, for it also obtained a grant from St. Cunégonde owing to its new location on the border between both municipalities. This advantage would seem to have been justified by the company's size, the number of employees (420 in 1888), and the value of its output. Such industry-friendly policies were undoubtedly factors weighing in the decision of Dominion Wadding to relocate in St. Cunégonde and of Colonial Bleaching & Printing Co. to move to St. Henri. Established in Sorel, Quebec, in 1883, Dominion Wadding agreed to move its factory to St. Cunégonde the following year in exchange for a 10-year abatement. Lachine made similar concessions to Dominion Bridge, which hesitated between setting up in this city or in Longueuil, on the south shore of the St. Lawrence. This industrial siting policy nevertheless came with a hefty price tag in the form of debt. As they were deprived of a major source of revenues, the industrial suburbs had to rely on only a portion of their potential tax base. Caught in the grip of severe financial hardship and towering debt, the new municipalities bordering Montreal (St. Henri, St. Gabriel, and St. Cunégonde) had no choice but to be annexed to the metropolis next door.

Distribution of Water in Montreal during a Shortage in 1913, Bibliothèque nationale du Québec, Michel Bazinet collection, CP 5141, Album 3-37-a, 1913

A TALE OF TWO CITIES

Water supply and the discharge of wastewater represented two priorities for outlying neighbourhoods then undergoing municipalization. At the time of its founding in 1875, St. Henri did not provide uniform water access to all of its residents. Its territory was crossed by a number of brooks, which had been an essential resource to local leather works, in addition to several wells. As a result, St. Henri's municipal authorities were highly disinclined to install a water supply system. Their attitude contrasted sharply with that of St. Cunégonde's officials, who, as soon as their municipality had been officially detached from St. Henri in 1876, awarded a contract for the construction of a water supply system. This signalled an apparent race to urbanize, as St. Henri, fearing to lose ground, in turn committed to building a water supply infrastructure three years later. It was only the first heat, however. On public health matters, St. Henri typically adopted a hands-off attitude. Even at the height of a smallpox epidemic in 1885, its municipal authorities hesitated to set up a health office. Thus it comes as little surprise that the town simply did not have a sewer system. Each year, a section of St. Henri was hit with floods, which in turn caused sewage water to overflow. In contrast with this approach, St. Cunégonde launched construction of its sewer system beginning in 1880. This network stood completed as of 1887, but it was only later that St. Henri started up its own infrastructures project. Against the backdrop of unsanitary environments typifying late-19th-century working-class localities, St. Cunégonde stood in marked contrast: there, "quality of life" seemed already to be an operative notion, although obviously the municipality had not yet attained the status of a garden city.

Clarke, F. R., *Unemployed Men Playing Ring Toss Game, Montreal*, McCord Museum, Notman Photographic Archives, MP-1988.59.59, ca 1935

COUPONS, CHEQUES, AND PUBLIC WORKS

For the people and industries located near the Lachine Canal, the 1929 stock market crash produced a variety of repercussions. Some industries were forced to lay off many of their employees. Others, such as Crane or Jenkins Valves, did not have to close as many jobs because they were able to find a ready outlet for their goods in the public infrastructure projects that were launched by several municipalities in order to counter unemployment. Far from being permanent, these programs were at best intermittent. When such public works were lacking, the jobless had to fall back on direct assistance payments that were defrayed in large part by municipalities and the Government of Quebec. However, as the Depression had deprived municipalities of a substantial share of their revenues, assistance eligibility criteria had to be tightened accordingly. First, unemployed workers had to have no other financial resources available to them, since by law, families were responsible for the welfare of any of their jobless members. Second, they had to have been a resident of the municipality for some time—a criterion that was reassessed depending on the depth of the economic turndown. In a municipality like Verdun, the jobless were provided relief measures that included the distribution of coupons, some of them redeemable at a single store and others bearing a given money amount. This system was quickly supplanted by allowances initially paid out in the form of cheques and then in cash. But although the unemployed of Verdun received benefits more generous than those offered in most other cities in Quebec, they nevertheless had to make do with amounts that kept their families far below the poverty line.

Montreal Light, Heat and Power Consolidated / LaSalle Gas,
Centre d'archives d'Hydro-Québec, Montreal Light, Heat and Power Consolidated collection (F9), F9 / 700763

"NOT IN MY BACKYARD"

On May 22, 1863, William J. Patterson, the grain board secretary, wrote to the Public Works Commissioners to complain that coal oil was being stored near flour warehouses and that the resulting odours risked spoiling quantities of flour. Patterson thus petitioned authorities to make it mandatory for petroleum products to be stored outside of the city limits—a request they assented to with respect to quantities in excess of five barrels. The canal authorities went even further, banning the transshipment of oil products beyond the Brewster basin in St. Cunégonde. Subsequently, oil imports from Ontario were to reach Montreal sporadically. In 1897, Imperial Oil established its tanks near the St. Henri railway bridge, in compliance with established zoning regulations, as housing had not been developed in this sector south of the canal. Sun Oil soon followed in its footsteps. Côte St. Paul went to all lengths to convey the image of an attractive town as a means of luring new taxpayers to its neighbourhoods—including the demolition, between 1889 and 1892, of all slaughterhouses and suet-, candle-, soap- and glue-works located within the municipality limits. During the first few decades of the 20th century, fertilizer companies such as Nutrite and Monsanto, not to mention LaSalle Gas, moved still further to the west, settling in as yet unurbanized areas. Nutrite, which had been set up by Canada Packers, used animal carcasses to manufacture its products, but according to its senior officers, the company had settled in LaSalle in order to take advantage of the sector's easy access to the greater Montreal road system.

Lachine Canal, View of St. Patrick and Grand Trunk Railway Bridges in Distance, Looking East, National Archives of Canada, Department of Transport collection, C 081529, 1907

A FAR CRY FROM THE GARDEN CITY

A municipality like St. Cunégonde covered an area of only 50 hectares. Within such narrow confines, it was not possible to lay out a great many parks or squares, all the more so because urban development there had been entrusted to speculators and private enterprise. In industrial communities, master plans were practically non-existent. Industries sought to settle along the edge of the canal for a variety of reasons: direct contact with transshipment facilities on the canal; access to a source of water that could not be delivered as cheaply by municipal water supply systems; and, finally, the possibility, at least for some, of returning wastewater into the canal, thus eliminating the cost of using municipal infrastructures. As for housing, single-family homes were uncommon in a landscape dominated by "plexes" of all shapes and sizes, regardless of the municipality they were located in. For the underlying principle of urban development remained the same for housing—making as much profit on lots as possible. That is why parcels became increasingly smaller and occupation densities higher, reaching a proportion of 35 residents per hectare in the early 20th century. St. Cunégonde had only one square to its name, whereas St. Gabriel was endowed with its first park, the St. Patrick Square, by the city of Montreal in 1899—12 years after its annexation. St. Henri could point to the George-Étienne Cartier Square, but only after its annexation in 1905. Although set up in the early 1910s, the Metropolitan Parks Commission was unable to implement its projects for want of adequate funding. In terms of actual area devoted to parks, Montreal was way behind Toronto and Boston. In the meantime, factory chimneys continued to spew their soot and fumes into the air.

Housing Conditions

Beaudoin Street, The First Home of the Lacasse Family in The Tin Flute, Archives nationales du Québec (Montreal), Conrad Poirier collection, P48, P11908, 1945

STAY PUT OR MOVE ON?

"Among the unstable inhabitants of the *ward*,* the annual moving crisis was already to be detected." This crisis was perpetuated in a quasi-ritual fashion in May of every year. Since the late 18th century, a majority of Montreal city-dwellers had been tenants, in keeping with the trend set in the city of Québec a few decades earlier. Neighbourhoods and municipalities located along the industrial corridor were no exception. While Montreal was in the throes of a housing crisis in the late 19th century, nothing indicates that the situation had evolved much 40 years later. The built environment varied considerably from one municipality or ward to another. St. Henri had preserved a number of small wood homes, but it also boasted brick-faced duplexes and triplexes of the kind encountered in St. Cunégonde. Housing conditions also differed from one municipality to another, although each bore its share of cramped, unsanitary, and overcrowded housing. Be this as it may, all municipalities were visited with a spring-time moving "itch" which saw tenants pull up stakes in hopes of finding better lodgings within the same neighbourhood. Although moving was a generalized phenomenon in Montreal and throughout urban Quebec, it is nevertheless difficult to determine exactly what proportion of tenants in fact moved. Some of the more privileged categories of workers were less mobile than others. Foremen, for the most part, occupied housing supplied or built by their employers. For a while, the Grand Trunk shops, Dominion Textile, and Dominion Bridge offered their foremen such amenities.

* *ward*: the ward referred to was that of St. Henri. Excerpted from Gabrielle Roy, *The Tin Flute*.

Arless, Sr., Richard, *Point St. Charles District, Montreal, P. Q.*, National Archives of Canada, Richard Arless collection, PA 043876, 1946

WORKERS' TENEMENT HOUSES

In an attempt to wake up Montreal businessmen to the needs of working-class families, the businessman, philanthropist, and city reform politician Herbert Brown Ames in 1896 published one of Canada's first sociological surveys, entitled *The City below the Hill*. The area he studied was bordered by McGill Street on the east, Dorchester Street (today's René-Lévesque Boulevard) on the north, the St. Lawrence River on the south, and the municipality of St. Cunégonde on the west. Among other issues, Ames was interested in housing conditions. Out of 4,709 buildings surveyed, 860 consisted in tenements (in French, *maisons de fond de cour*—literally, "back court houses")—that is, 18% of the total dwellings, accounting for 10% of all households in the sector. Blaming these habitations for all manner of ills, he purely and simply wanted to eliminate them. Ames was unable to make much headway in his crusade since, 40 years later, a survey of housing conditions in the Montreal neighbourhood bordering the canal listed 2,779 such buildings encompassing 4,007 apartments, of which more than a third were considered to be in only fair condition, if not outright poor condition. On the other hand, windowless rooms now appeared to have become a thing of the past, accounting for only 18 of the housing units surveyed. As a rule, a family of four lived in a four-room apartment. Thus, it may be inferred that duplexes were the most widespread type of tenement housing and that overcrowding was something of a marginal phenomenon in these working-class dwellings. In fact, duplexes were most often encountered in the older of the wards (including those lying close to the canal), where lots were widest. After 1880, they usually measured 8 metres across the front and no longer afforded the room necessary to pass into the yard or alley behind. It is also worth noting that this state of overcrowding fluctuated little from the 1870s on, at least in the Francophone sectors.

The Standard, *Typical Urban Scene in Montreal: Tenements*, Bibliothèque nationale du Québec, E.-Z. Massicotte albums, 1914

WORKING-CLASS NEIGHBOURHOODS AND PUBLIC HEALTH

Although the Government of Quebec adopted its *Public Health Act* in 1886, the impact of this legislation in working-class neighbourhoods was mitigated to say the least. In Montreal, infant and child mortality was the most critical problem confronting health authorities. In the late 19th century, one out of three babies died before the age of one (a statistic close to the situation prevailing in the city of Québec during the 18th century, when two out of five children died before they had turned one). Montreal's infant mortality rate was third highest worldwide, with peak rates occurring in the neighbourhoods bordering the Lachine Canal. In addition, many children were afflicted with respiratory diseases, and adults grappled with tuberculosis, the second leading cause of death in Montreal. Tuberculosis infection rates varied according to age and gender, but the etiology of this disease also embraced a range of environmental factors such as diet, overcrowding, and working conditions. Obvious examples of workplace-related causes include factories where dust pollution levels were high, such as textile plants and large-scale flour mills. Being a respiratory illness, tuberculosis was only aggravated by living in a soot-ridden environment. And, being a contagious disease, tuberculosis spread more readily in crowded settings, such as cramped living quarters, congested and poorly ventilated factories, or packed classrooms offering 2 cubic metres of air per student, instead of the 7-cubic-metre norm targeted by Swedish and British legislators and which Quebec authorities wished to use as a basis for school and workplace standards. What, then, is one to think of the 17 cubic metres allotted British soldiers by the army staff in the mid-19th century?—No doubt about it, city living in a manufacturing environment came with a hefty price tag.

THE WITNESS, *Child Welfare Exhibition*,
Bibliothèque nationale du Québec,
E.-Z. Massicotte albums, 1912

THE CHILD WELFARE EXHIBITION OF 1912

Imbued with idealism and inspired by late-19th-century philanthropic currents, this exhibition was devoted to everything that directly or indirectly affected the social and physical condition of children. Starting from the postulate that "it is the man that makes the nation, it is the child that makes the man," the event's promoters had the stated objective of "training public opinion." Thus, the exhibition took up such issues as housing, environment, education, religion, and poverty—not to mention the care to be provided child. Based on a typical household budget worked up for the purposes of this show, an unskilled worker could earn approximately $550 if he worked an entire year. He then would earmark 20% of his income for rent, 50% for food for himself and his family, 5% for heating, 13% for clothing, 4% for lighting, 4% for city transportation, 1% for water tax expenses, and 2% for unforeseen circumstances. This scenario obviously accounted neither for inflation nor for illness of the family provider, in which case mother and child might have to pitch in, even if the child was still under 14. Indeed, necessity was the mother of—*subvention.* Thus commenced the infernal cycle of poverty, since such meagre wages were inadequate to afford the working man a decent apartment, much less to educate his children properly—thus bringing the circle of illness and unskilled labour relentlessly back round on itself. Obviously, the airless, sunless workers' slums were singled out as sources of tuberculosis and infant mortality. Herbert Ames ultimately succeeded in arousing concern for the plight of working-class families among the high-minded. Strangely enough, however, the promoters of the exhibit had little to say about what was required to remedy the situation.

C. A. Barbier photography studio, *The Ice Cutters*, Photography collection of the Musée de Lachine, ca 1900

ICE, A NECESSARY FOOD SUPPLEMENT

There are several methods for preserving food, involving primarily salting, canning, or refrigerating. From the beginning of the colony, ice was used to conserve food. In an era when households were not yet equipped with our modern refrigerators and freezers, an icebox did the job. Occasionally installed on the rear balcony, in winter as in summer, iceboxes became particularly popular as a home appliance in the years following World War I. For mothers unable to breastfeed, fresh milk was an essential replacement food for babies. As living standards rose between 1900 and 1920, iceboxes were relied on to improve milk conservation. In Montreal during the first two decades of the 20th century, industrial milk was the source of numerous infantile health problems, particularly in connection with unpasteurized milk. In this context, the refrigeration of milk proved to be an essential measure, at least for families that were unable to benefit from the programs of organizations such as the "Goutte de lait," which distributed pure fresh milk throughout the disadvantaged neighbourhoods of the city. Thus, every winter, blocks of ice were harvested from the Lachine Canal, the Lachine basin in particular, and stored in sawdust-insulated warehouses for use during hot summer days.

Urban Mass Transit

Wellington Bridge on the Lachine Canal,
Port of Montreal Archives, photo 1623, 1910

NEXT DOOR NEIGHBOUR TO THE FACTORY

Beginning in the mid-19th century, a rail connection was established between Montreal and Lachine. In 1861, the urban transit system of Montreal came into being with a streetcar service, horse-driven at first but electrified in 1892. Taken together, do these developments point to the conclusion that throngs of workers boarded trams to get to and from work? Any answer to this question involves ascertaining the cost and number of trips recorded by the transportation company, knowing the main itineraries offered, but above all determining where workers lived in relationship to their place of employment. Whereas in 1863, the Montreal City Passenger Railway recorded one million trips, at a time when the city's population barely attained 120,000, in 1911 the Montreal Street Railway had a ridership of more than 118 million, for a population of slightly more than 550,000. Excepting a slowdown during the Great Depression, this trend continued to progress until the end of World War II, when yearly ridership settled at almost 400 million. In a parallel development, the system was extended to outlying cities such as Verdun as of 1899, whereas routes had already existed for some time on both sides of the Lachine Canal. In 1910, the fare for a single trip was $0.05, which would seem quite inexpensive. It should be remembered, however, that an unskilled worker earned $550 per annum during this time and that tramway travel took up 4 to 5 % of his income—at minimum, considering that this proportion holds for a period of full employment, hence of full pay. However, a survey conducted in 1937 by Imperial Tobacco of St. Henri on the place of residence of its employees has much to say about the daily travels of its workers: out of a total of close to 1,600 employees, slightly more than half (53%) inhabited less than a kilometre and a half from their place of employment, and 76% of all employees resided less than three kilometres from the factory. There are strong grounds for thinking that the situation was little different elsewhere. In short, the proximity of workplace to home had scarcely changed since the mid-19th century.

Repaired South Wall above Smith Lock No. 5, Looking East, Lachine Canal, National Archives of Canada, Department of Public Works collection, PA 110032, 1928

THIS TOO SHALL COME TO PASS...

In 1824, only 13 barges and batteaux travelled up the portion of the canal system open at that time. The following year, after the canal was made fully open to navigation, the number of transits increased to 727 and mounted steadily thereafter. In 1840, more than 4,000 vessels of all kinds were recorded as moving through the canal, a figure that works out to approximately 19 boats a day. This spectacular increase in traffic prompted the authorities to propose widening the canal. In 1873, on the eve of the second widening, more than 11,000 boats used the canal system—that is, an average 51 per day. This figure declined slightly during the 20th century, when an average of close to 8,800 ships transited through the canal annually. Paradoxically, this average declined yet further following the extension of the shipping season in 1913 that year, only 8,600 vessels used the canal, for a nonetheless respectable average of 37 a day. In 1960, after a portion of the canal had been closed to traffic, nearly 2,000 boats were still recorded on the canal. However, this figure would continually erode, dropping to less than 200 annually between 1965 and 1968. Posting a total of 13,572, the year 1870 stands out for the greatest number of boats travelling through the canal. Such statistics provide insight into scenes of the kind depicted in this photo, where road bridges over the canal can be seen closed to drivers and pedestrians, who stand patiently by as a boat completes its passage.

Federal Model X. Delivery Truck,
Redpath Sugar Archives, ca 1919

THE TRUCKING ERA BEGINS

"Corridor" is a term frequently-used in connection with the Lachine Canal, variously denoting the shipping lane, the neighbouring railway network, and the one-time embryonic regional road system which transformed into a full-fledged highway grid. The railway line along the canal was part of a main link between Montreal and Toronto. Beginning in the late 1920s, trucking was increasingly looked to as a means of hauling wares from one end of the city to the other or of carrying goods between factories. As trucks offered greater mobility, they soon became the preferred mode of urban transportation among major Lachine Canal manufacturers, such as Canada Sugar (Redpath Sugar) or Dominion Bridge. For the smaller machined parts turned out in its industrial complex, Dominion Bridge readily relied on trucking, since this solution enabled it to deliver its products to the most remote locations that had no ship or train service. In a particularly telling comment, one of the senior executives of Nutrite, a subsidiary of Canada Packers, noted that when the company located to Ville LaSalle, the distance from the central business district was not viewed as a handicap to its activities, since it relied on trucking for its shipping needs.

Daily Life around the Canal

Gabrielle Roy and Boys near St. Ambroise Street Crossing,
Archives nationales du Québec, Conrad Poirier
collection, P48, P11917, 1945

GABRIELLE ROY'S ST. HENRI

Bonheur d'occasion (called *The Tin Flute* in English) was Gabrielle Roy's first novel, and it met with international acclaim from the time of its publication, first in Montreal in 1945, then following a second printing in Paris. It won the prize of the nascent Académie canadienne-française as well as the Prix Fémina, one of France's most coveted literary awards. Film rights to the work were sold to a Hollywood producer, and the novel was translated into English in 1947 and, eventually, seven other languages. The success of this novel of everyday life in Montreal was very likely related to the way it broke with the French-Canadian literature created until that time. Several of its protagonists were unskilled workers, overwhelmed by the dire economic straits forced on them by the Great Depression. Some of these recently urbanized characters had few resources with which to extricate themselves from the working-class slums in which the novel was set. However, all lived with the hope of one day escaping to a better life. For some, like Azarius and Eugène, father and brother of the novel's heroine, Florentine, the way out was war and the potential monetary and material advantages to be had by enlisting. There was no small irony involved in this alternative, as the war years also enabled the factories of the canal corridor and St. Henri to prosper once again. The power of this novel also derives from Gabrielle Roy's uncannily realistic descriptions of the landscape, home interiors, and the confusion of moving day, all of which were the result, according to the novelist, of careful field observations of a quasi-sociological nature. Such details thoroughly and convincingly immerse readers in the environment and mentality of pre-war St. Henri.

Youth Diving into the Lachine Canal, St. Henri Historical Society, 23.1-5.4, ca 1940

CANAL BATHING—NOT FOR THE FAINT OF HEART!

Although the discharging of industrial wastewater was regulated to a certain degree, the ordinances were not always strictly enforced. To be sure, starting from the mid-19th century, sawmills were prohibited from dumping sawdust into the canal. However, this ban, an integral condition of the lease binding the canal administration and a given sawmill, was not intended to protect water quality but instead to prevent water intakes from clogging. And only a few leases explicitly prohibited the effluents of tanneries, slaughterhouses, and chemical factories from entering the water. In the case of the Montreal Stock Yards, the provisions of its lease running from 1898 to 1928 were perfectly clear: the company was forbidden from discharging animal remains or any other matter that might adversely affect "water purity." At best, other leases simply stipulated that any wastewater could be discharged into the canal "if it was not contaminated." Furthermore, shipping regulations contained no provisions concerning hazardous spills. Perhaps this situation owes to the fact that diesel engines were very little used prior to the end of World War II. All in all, it comes as no surprise that swimming in the canal was banned by regulation. However, there are grounds for wondering whether it was environmental or safety concerns that weighed heaviest in this decision. Whatever the reason, dog days nevertheless prompted the more intrepid of youth to take the plunge.

Epilogue

THIS WORK could have been more prosaically titled "Managing Water" or "Controlling the Water Resource," for had it not been for water, there would not have been a canal, or ships for that matter, and few factories would have likely been erected here. Indeed, it is fair to assume that the siting of industries was directly related to the three fold use made of the Lachine Canal. Its function as a shipping lane is obvious, and that of energy source has often been referred to. On the other hand, its role as a supply of water to industries is less well known. Whatever the case, water and, above all, its management was the source of numerous clashes among canal users.

It could scarcely have been otherwise, since water in the view of the industrial sector represented a resource to be utilized, whereas for the shipping sector, it was a means of navigation. With its unquenchable thirst for water, industry disturbed the already limited water depth required by transiting vessels. In other words, whenever factories siphoned away water for their industrial needs until the level of a reach was lowered by several centimetres, canal-bound ships, especially those that had been loaded to full capacity and used all available depth, experienced enormous difficulty negotiating their way through the system. This phenomenon, coupled with the silt-laden currents responsible for creating shoals, increased the risks of running aground among ships having a deep draught. And that was but one of the water management problems confronting canal engineers.

Industries that drew on canal water to supply their energy needs required a rapid current, in utter contrast to shippers, who needed a slower current, particularly when they sailed upstream. To further complicate the situation, water from hydraulic power facilities was released into the canal and the resulting turbulence made lock docking or exiting manoeuvres perilous. Viewed in this light, many of the pilots' recriminations seem justified. The canal water conundrum persisted primarily throughout the second half of the 19th century and headed up the list of obstacles which engineers first tackled. They generally adopted piecemeal solutions in an attempt to comply with the demands of shipowners and industrialists alike. After installing or widening weirs up and down the canal—that is, after increasing the intake of water—they resigned themselves to widening the canal, a decision which afforded the canal sector a certain degree of stability during the 20th century.

The availability of canal water weighed heavily in the decision of some industries to settle in the sector—to the same extent as the availability of lots or "environmental" norms, for example. Whereas the municipal water supply only just barely provided the area's citizens with 350 litres of water per day, canal factories consumed water by the millions of litres. For these manufacturing firms, the location of lots was very much a secondary consideration in comparison with the possibility of tapping into an abundant supply of water. However, the mass of water to be had in the reaches of the oldest industrial areas appeared unequal to the insatiable needs of some factories. It was all the more inadequate since the water in these reaches had to be shared among several industries that even then were pumping unreasonably large volumes. That is why some manufacturing firms having gargantuan appetites chose to establish plants in the unoccupied portions of the corridor that also corresponded to the longest reach. The conflict between shipping and manufacturing interests militated in favour of this decision, for had it been only an issue of lots, nothing would have prevented industries from settling in the oldest sections and purchasing the land they needed. But the reach would have been sucked dry had they done so. A fragile coexistence reigned between the shipping and manufacturing sectors, and to a certain extent it shaped the development of the corridor and its surroundings.

This landscape was the product not only of the industries located in it, but also of the very mission to which the canal was dedicated. While early in the history of the canal, Montreal's merchants had intended to compete with New York harbour as a grain-handling port, it took them many many years before they were able to accomplish their objective. Montreal became the leading grain-handling port only in the 20th century, a development that nevertheless left its mark on the canal profile. Together, the wheat boom and the final standardization of the St. Lawrence canals in the early 20th century triggered the construction of grain elevators at the Montreal end of the canal. (It should be noted, however, that the function of grain transshipment was assumed by the sector beginning in the mid-19th century.) In response to this trend, the canal system administrators decided to deepen the first reaches of the canal in order to facilitate access by ocean-going ships. As a result, this portion of the canal became an interface between inland and trans-Atlantic navigation.

Inland shipping depended on boats known as "canallers," specially designed for canal traffic. They carried bulk—wheat, coal, and lumber, for the most part. In the late 19th century, the shipment of wheat gave rise to a partnership between the industrial and marine transportation sectors within one of the country's largest shipping firms, the Montreal Transportation Co. In turn, this concern was incorporated within Canada's leading shipper, Canada Steamship Lines, which had itself resulted from the merger of several shipping companies. The trend toward concentration was by no means specific to the marine transportation sector, since during the same period numerous manufacturing firms located near the canal also merged to form Canadian-wide conglomerates.

The industries established along the Lachine Canal were diversified, to say the least. While it is true that Montrealers at one point dubbed the canal valley "Little Lowell," it should be noted from the outset that the canal's industrial corridor considerably outstripped the Massachusetts-based manufacturing complex in terms of the actual numbers of firms and the diversity of their output. Lowell was concentrated in textile manufacturing, whereas the canal's factories produced just about everything, from sugar to rubber, with chemicals and steel in between. In fact, from the very beginning, manufacturing in the area was not confined to a specific sector, as heavy and light industry lived cheek by jowl along the canal banks.

More than 600 companies occupied factories along the edges of the canal at one time or another until the end of World War II. Obviously, these firms chose to locate here on account of the ease of transportation, since the canal served not only as a marine shipping terminus but also as the hub of the nation's railway system, following the establishment of the Grand Trunk shops in Point St. Charles during the second half of the 19th century. The critical mass embodied by these industries also produced a certain power of attraction on other companies that ultimately chose to settle along the canal. For a frequently ignored facet of industrial activity is that manufacturing firms not only produce for the domestic market, they also turn out goods for sale among themselves.

Thus, for example, the siting of the Grand Trunk Railway shops in Point St. Charles had numerous repercussions for the Lachine Canal industrial complex. For one, it led to the establishment of wheel or car suspension production units, as well as nail works specializing in the production of tacks for upholstering passenger cars, or spikes for solidifying tracks. Locomotives were fitted out with brass bells, also produced by canal factories. Locomotives and cars were made of iron and steel and had to withstand the rigours of the Canadian climate—and, with this, rust. Inevitably, they would have to be painted, and repainted. There is no want of examples in connection with rail transportation building and repair.

Carbochemistry represented another of these poles of industrial development, as was shown by the establishment of three gas manufacturing complexes along the banks of the canal. In addition to producing lighting gas, these gas plants also produced coal gas and, in the process, a long list of by-products that were highly useful to other industries along the canal (although manufactured gas continued to be their main product). However, with the advent of electricity during the 1880s, the energy picture changed, and these companies now had to produce gas for heating and cooking. In the 1920s came a new addition to the product line: metallurgical coke, a prized fuel source in heavy industry. Another example of the closely interknit relationships between companies was mauveine (otherwise known as Perkin's violet), the first commercial pigment synthesized from coal-tar intermediates, whose discovery breathed new life into the dye industry during the 1850s. The Lachine Canal manufacturing complex would also be home to several factories specializing in textile dyeing. And, in a particularly elaborate end use, cotton goods entered into the production of electric cables, thus creating a new outlet for the textile sector during the 20th century.

The emergence of a critical mass of factories along the canal prompted workers to settle in the vicinity of their workplace. More than 20% of Montreal Island's workforce held jobs in the sector. Thus, as a corollary to this industrial development, southwest Montreal Island entered a period of urbanization. Local authorities attempted to woo these industries to settle in their respective municipalities and liberally granted them long-term tax breaks and a sprinkling of other perquisites. Wages constituted another factor that caused workers to cluster within close range of the factory gates. A combination of low pay and non-existent mass transit condemned them to living in often substandard housing, in surroundings made grey by the soot sent skyward from factory chimneys day after day. In itself, this airborne pollution was indicative of another phenomenon stemming from the marine environment of the canal, for industrialists took advantage of the waterway to receive coal from three different sources—Nova Scotia, Pennsylvania, and Great Britain. So doing, they were able to play the coal wholesalers off of one another in order to keep their energy-related expenses relatively low. From once having been "Little Lowell," the canal now became "Smoky Valley."

Exacting working conditions, low wages, malnourishment, inadequate housing, unhealthy ambient air—all such factors clearly serve to explain why the neighbourhoods and wards bordering the canal claimed the dubious distinction of abnormally high mortality rates and record-high tuberculosis infection rates. World War I nevertheless forced industrialists to increase their use of electricity—a changeover that helped make the environment somewhat cleaner. Furthermore, the wheat boom, combined with the development of western Canada and the outbreak of World War I, contributed to improving the economic conditions of workers, with the result that on the eve of World War II, the situation in the areas bordering the canal had improved, at least in terms of public health.

Ironically enough, the use of electricity, which portended the sweeping changes shortly to occur in Quebec's industrial landscape, also heralded the end of the golden age of Montreal's manufacturing industry. A wave of hydroelectricity-driven industrialization was about to unfurl over Quebec (with paper mills and aluminum mills in the forefront), but no major pulp or aluminum plant chose to set up operations in Montreal or along the edges of the canal. Although at the time such signs appeared inconsequential from the viewpoint of the country's leading industrial city, they nevertheless foretold of years—the "electric era"—that would hold out less promise for Montreal than the "steam era."

The coming years would be all the less promising since the hydro rate schedule applied in Montreal during the interwar period placed the city at a disadvantage relative to Canada's other industrial regions, and above all Toronto, which was then embarking on a period of full expansion. Beginning in 1906, Ontario made a move that had major import for future events: it nationalized hydroelectricity as a means of leveraging the province's industrial development. Quebec opted for the opposite course of action, leaving the development of hydroelectricity to private enterprise. As it so happened, Montreal Light, Heat and Power held a monopoly on electricity in Montreal, and in contrast with its rival, the Shawinigan Water and Power Company, it did not offer preferred rates to the industrial sector. Given the absence of such a policy, MLHP ended up redirecting industries in quest of cheap energy to other, more clement environments.

The Lachine Canal and its industrial complex provide a window not only onto the history of Montreal and Quebec but also onto that of modern Canada. Only just beneath the surface of these events, it is possible to make out the evolution of inland shipping and a portion of the Canadian merchant marine, the development of a manufacturing industry, the growth of the labour movement, and the passage from a rural to an urban world.

Select Bibliography

The titles appearing below are only a partial selection of our general bibliography. Some omissions may have occurred owing to the great number of works involved.

Studies and Theses

Commercial Sketch of Montreal and its Superiority as a Wholesale Market, Montreal: Chisholm & Dodd Publishers, 1868

Ames, Herbert Brown, *The City below the Hill: A Sociological Study of a Portion of the City of Montreal, Canada*, with an introduction by P.F.W. Rutherford,Toronto: University of Toronto Press, 1972 (1897)

Bischoff, Pierre, *Tensions et solidarité: la formation des traditions syndicales chez les mouleurs de Montréal, Hamilton et Toronto, 1851 à 1893*, Ph.D. thesis in history, Université de Montréal, 1992

Borthwick, John Douglas, *Montreal, Its History, to Which is Added Biographical Sketches, with Photographs, of Many of its Principal Citizens*, Montreal: Drysdale and Co., 1875

Bradbury, Bettina, *Working Families: Age, Gender and Daily Survival in Industrializing Montreal*, Toronto: McClelland and Stewart, 1993

Camu, Pierre, *Le Saint-Laurent et les Grands Lacs au temps de la voile, 1608-1850*, LaSalle (Quebec): Hurtubise HMH, Cahiers du Québec (collection Géographie), 1996

Carey, Luc, *Un type de maison ouvrière: le déclin de la maison de fond de cour à Montréal, 1880-1920*, M.A. thesis in history, Université du Québec à Montréal, 1996

Charland, Jean-Pierre, *Les pâtes et papiers au Québec, 1880-1980. Technologies, travail et travailleurs*, Québec: Institut québécois de recherche sur la culture, 1990

Clavette, Susanne, *Des bons aux chèques: aide aux chômeurs et crise des années 1930 à Verdun*, M.A. thesis in history, Université du Québec à Montréal, 1986

Collard, Edgar Andrew, *Passage to the Sea: The Story of Canada Steamship Lines*, Toronto: Doubleday Canada, 1991

Copp, Terry, *The Anatomy of Poverty: The Condition of the Working Class in Montreal, 1897-1929*, Toronto: McClelland and Stewart, 1974

Daumas, Maurice (editor), *A History of Technology & Invention; Progress through the Ages, Vol. 3. The expansion of mechanization*. Translated of *Histoire générale des techniques. L'expansion du machinisme*, by Eileen B. Hennessy, New York: Crown Publishers, 1970

Feltoe, Richard, *Redpath. The History of a Sugar House*, Toronto: Natural Heritage/Natural History Inc., 1991

Feltoe, Richard, *Let Redpath Sweeten It*, Toronto: Natural Heritage/Natural History Inc., 1993

Hadfield, Charles, *British Canals, An Illustrated History*, Vancouver : David & Charles, 8th edition, 1992

Harvey, Fernand, *Le mouvement ouvrier au Québec*, Montreal: Boréal Express, 1980

Harvey, Fernand, *Révolution industrielle et travailleurs. Une enquête sur les rapports entre le capital et le travail, à la fin du 19e siècle*, Montreal: Boréal Express, 1978

Heisler John P., *The Canals of Canada*, Ottawa: Department of Indian Affairs and Northern Development, Collection Canadian Historic Sites: Occasional Papers in Archaeology and History, No. 8, 1973

Kilbourn, William, *The Elements Combined: A History of the Steel Company of Canada*, Toronto: Clarke Irwin & Co. Ltd., 1960

Lasserre, Jean-Claude, *Le Saint-Laurent, grande porte de l'Amérique*, LaSalle (Quebec): Hurtubise HMH, Cahiers du Québec (collection Géographie), 1980

Lauzon, Gilles, *Habitat ouvrier et révolution industrielle: le cas du village Saint-Augustin*, Montreal: Regroupement des chercheurs et chercheures en histoire des travailleurs, 1989

Lavigne, Marie, and Yolande Pinard, *Travailleuses et féministes. Les femmes dans la société québécoise*, Montreal: Boréal Express, 1983

Lewis, Robert, *Manufacturing Montreal. The Making of an Industrial Landscape, 1850 to 1930*, Baltimore: John Hopkins University Press, 2000

Lewis, Walter, *Until Further Notice: The Royal Mail Line and the Steamboat Trade of Lake Ontario and the Upper St. Lawrence River, 1837-1875*, M.A. thesis, Queen's University, 1983

Linteau, Paul-André, *Histoire de Montréal depuis la Confédération*, Montreal: Boréal, 1992

Lippel, Katherine, *Le droit des accidentés du travail à une indemnité: analyse historique et critique*, Master of Law thesis, Université de Montréal Faculty of Law, 1986

McCullough, Alan Bruce, *The Primary Textile Industry in Canada: History and Heritage*, Ottawa: Canadian Parks Service, National Historic Sites, 1992

Merrilees, Andrew, *The Railway Rolling Stock Industry in Canada. A History of 110 Years of Canadian Railway Car Building*, Toronto: Andrew Merrilees Limited and Merrilees Equipment Limited, 1963

Pinon, Pierre, *Canaux: rivières des hommes*, Paris: Desclée de Brouwer, 1995

Price, Enid Margaret, *Changes in the Industrial Occupations of Women in the Environment of Montreal during the Period of the War, 1914-1918*, Montreal: Canadian Reconstruction Association, 1919

Rens, Jean-Guy, *The Invisible Empire: A History of the Telecommunications Industry in Canada*, (translated by Käthe Roth), Montreal: McGill-Queen's University Press, 2001

Ricard, François, *Gabrielle Roy*, Montreal: Fides, 1975

Richelieu & Ontario Navigation Co., *The Beauty Spots of Canada. Descriptive of that Delightful Trip down the River St. Lawrence and up the World-Famed Saguenay*, Montreal: Desbarats & Co., 1895

Robert, Jean-Claude, *Atlas historique de Montréal*, Montreal: Art global, Libre Expression, 1994

Roquigny, Peggy, *Sainte-Cunégonde: société et territoire dans un contexte d'industrialisation, 1876-1905*, M.A. thesis in history, Université du Havre (France) and Université du Québec à Montréal, 1999

Rouillard, Jacques, *Histoire du syndicalisme au Québec. Des origines à nos jours*, Montreal: Boréal, 1989

Rouillard, Jacques, *Les travailleurs du coton au Québec 1900-1915*, Montreal: Les Presses de l'Université du Québec, 1974

Roy, Gabrielle, *The Tin Flute* [Translation of *Bonheur d'occasion*, by Hannah Josephson, Montreal: McClelland and Stewart Limited (New Canadian Library no. 5), 1947; reprinted in 1964

Salmon, Stephen, "A Prosperous Season": Investment in Canadian Great Lakes Shipping, 1900-1914, in Victoria Brehm, *"A Fully Accredited Ocean": Essays on the Great Lakes*, Ann Arbor: The University of Michigan Press, 1998

Shaw, Ronald E., *Canals for a Nation, The Canal Era in the United States, 1790-1860*, Lexington: University Press of Kentucky, 1990

Stevens, George Roy, *The Belding-Corticelli Story, 1876-1951*, Montreal: Belding-Corticelli, 1951

Tulchinsky, Gerald Joseph Jacob, *Évolution des moyens de transport dans la région du Saint-Laurent et des Grands Lacs, 1828-1860 [slides] / Transportation Changes in the St. Lawrence-Great Lakes Region, 1828-1860*, Ottawa: Canadian Museum of Civilization / National Film Board of Canada, 197

Tulchinsky, Gerald Joseph Jacob, *The River Barons: Montreal Businessmen and the Growth of Industry and Transportation, 1837-53*, Toronto: University of Toronto Press, 1977

Way, Peter, *Common Labor. Workers and the Digging of North American Canals, 1780-1860*, Baltimore: John Hopkins University Press, 1997

Wilson, George H., *The Application of Steam to St. Lawrence Valley Navigation, 1809-1840*, M.A. thesis in history, McGill University, 1961

Reports

Archemi et al., *Inventaire et évaluation des ressources culturelles du canal de Lachine*, Montréal: Report submitted to Parks Canada, 1995, 3 vol.

Bernier-Morel, Lyne, Y. Desloges, C. Piché, and G. Piedlue, *Le canal de Lachine et ses usines : témoins de l'industrialisation canadienne*, Québec: Report submitted to the Historic Sites and Monuments Board of Canada, Parks Canada, April 1996

Contant, Isabelle, *Évolution du trafic du canal de Lachine de 1825 à 1873*, Québec: Canadian Parks Service, 1983

Espesset, Hélène, *Canal Lachine: dossiers structuraux*, Québec: Parks Canada, 1978-1979

Gelly, Alain, *De l'eau et de la fumée. Forces motrices au canal de Lachine 1846-1940*, Québec: Parks Canada, 2001

Gelly, Alain, *L'importance du Montréal industriel dans l'ensemble canadien*, Québec: Parks Canada, 1999

Gelly, Alain, *Production et fabrication pré-industrielles et industrielles: cinq productions associées au canal de Lachine*, Québec: Report submitted to Parks Canada, 1998

Lafrenière, Normand, *Évolution du trafic sur le canal de Lachine et les canaux du St-Laurent de 1870 à 1920*, Québec: Canadian Parks Service, 1985

McNally, Larry, *Water Power on the Lachine Canal, 1846-1900*, Québec: Parks Canada, 1982

Sévigny, André, *L'urbanisation dans le corridor du canal de Lachine dans la deuxième moitié du 19e siècle*, Québec: Parks Canada, 1986, 2 vol.

Willis, John, *The Process of Hydraulic Industrialization on the Lachine Canal, 1840-1880: Origins, Rise and Fall*, Québec: Canadian Parks Service, 1987

Articles

Dubuc, Alfred, "Montréal et les débuts de la navigation à vapeur sur le Saint-Laurent," *Revue d'histoire économique et sociale*, vol. XLV, 1967, pp.104-121

Lauzon, Gilles, "Cohabitation et déménagements en milieu ouvrier montréalais; essai de réinterprétation à partir du cas du village Saint-Augustin (1871-1881)," *Revue d'histoire de l'Amérique française*, vol. XLVI, no. 1 (summer 1992), pp. 115-142

Lewis, Walter, "The Canadian Navigation Company, 1861-1875," *Fresh Water*, vol. 1, no. 1 (spring 1986), pp. 4-14

McNally, Larry, "The Relationship between Transportation and Water Power on the Lachine Canal in the Nineteenth Century" in Richard A. Jarrell and Arnold E. Roos, *Critical Issues in the History of Canadian Science, Technology and Medicine / Problèmes cruciaux de l'histoire de la science, de la technologie et de la médecine au Canada*, Ottawa: HSTC Publications, 1981, pp. 76-88

Neilson, Rick, "The Montreal Transportation Company," *Fresh Water*, vol. 4, no. 1, 1989, pp. 21-24

Salmon, Stephen, "Fool Propositions": Mackenzie King, The Dominion Marine Association and the Inland Water Freight Rates Act of 1923, *Fresh Water*, vol. 2, no. 1 (summer 1987), pp. 8-19

Salmon, Stephen, "This Unsatisfactory Condition": The Formation and Financing of Canada Steamship Lines, 1910-1915, *International Journal of Maritime History*, vol. XII, no. 1 (June 2000), pp. 145-176

Tulchinsky, Gerald Joseph Jacob, "Une entreprise maritime canadienne-française. La compagnie du Richelieu, 1845-1854," *Revue d'histoire de l'Amérique française*, vol. 26, no. 4, (March 1973), pp. 559-583

TYPESET IN SCALA 10 BY JOSÉE LALANCETTE,
THIS BOOK WAS PRINTED IN JUNE 2002
AT MARC VEILLEUX IMPRIMEUR IN BOUCHERVILLE
FOR DENIS VAUGEOIS, PUBLISHER OF SEPTENTRION